Biometrics: A Very Short Introduction

VERY SHORT INTRODUCTIONS are for anyone wanting a stimulating and accessible way into a new subject. They are written by experts, and have been translated into more than 45 different languages.

The series began in 1995, and now covers a wide variety of topics in every discipline. The VSI library currently contains over 550 volumes—a Very Short Introduction to everything from Psychology and Philosophy of Science to American History and Relativity—and continues to grow in every subject area.

Very Short Introductions available now:

Available soon:

Michael Fairhurst

BIOMETRICS

A Very Short Introduction

OXFORD
UNIVERSITY PRESS

OXFORD

UNIVERSITY PRESS

Great Clarendon Street, Oxford, OX2 6DP,
United Kingdom

Oxford University Press is a department of the University of Oxford.
It furthers the University's objective of excellence in research, scholarship,
and education by publishing worldwide. Oxford is a registered trade mark of
Oxford University Press in the UK and in certain other countries

First edition published in 2018

Impression: 1

Published in the United States of America by Oxford University Press
198 Madison Avenue, New York, NY 10016, United States of America

British Library Cataloguing in Publication Data
Data available

Library of Congress Control Number: 2018950715

ISBN 978-0-19-880910-4

Printed in Great Britain by
Ashford Colour Press Ltd, Gosport, Hampshire

Contents

Preface

It cannot have escaped anyone's notice that questions about establishing and monitoring personal identity have increased in importance with the rise in the availability and ubiquity of automated processing systems and powerful information platforms. Many of our day-to-day interactions are now conducted via smartphones, tablets, and other computer-based systems and, even in interpersonal transactions which are undertaken more directly, we have become, of necessity, more concerned with issues of security. More often than not, this is likely to involve situations in which we need to have trust and confidence that people really are who they claim to be.

The tools which technology provides to help us in this task of identifying ourselves to others, and others to ourselves, have developed significantly over the years, even though we still often routinely encounter systems and processes which we know have been around a long time, and the vulnerabilities of which are (sometimes acutely) obvious. Access to our physical spaces—house, office, car, and so on—is often still protected by the use of a physical lock and key (although we now also frequently add a second protective layer involving some sort of alarm system) and, despite ongoing developments and a wider variety of options, the use of 'chip and PIN' strategies for securing, not just physical spaces, but also access to important data (our bank accounts, for example) defines a process with which very few readers will be unfamiliar.

Of course, we live in the age of the computer (we often use the term 'information platform' nowadays, given the variety of powerful computing devices to which we have access, and their vast and diverse functionality. Indeed, the computational power of a regular smartphone is staggering to anyone who remembers early home computer systems), and we are now highly dependent on an ability to access online a huge range of data and services. An implication of this is that it has become more important than ever to be able to demonstrate that we, as system users, have the authority to access certain 'virtual' as well as physical spaces, while ensuring others do not.

It is changes such as these which have, if not initiated, then certainly stimulated, the development of a new and fundamentally different method of establishing or confirming the identity of individuals. In this context 'biometrics' refers to an approach which moves away from physical or abstract tokens for this purpose, and which instead uses objectively measurable characteristics of each individual to determine identity. Since we carry with us, all the time, our fingerprint (to take just one example), we no longer need to worry about losing a key, or forgetting a password, and the bond between an individual and a demonstrable identity becomes stronger and more secure.

This book introduces the principal concepts of biometrics, from an understanding of how these relatively new technologies emerged, the principles on which they function, how recent developments have improved both their reliability and their applicability, to where the whole biometrics enterprise might be heading. The book aims to take the reader on a journey to reveal the important features of the past, present, and future for this important, powerful, and highly effective approach to addressing some of the most challenging issues confronting modern society.

List of illustrations

Chapter 1
Are you who you say you are?

Introduction

We live in a society which is increasingly interconnected, where communication between different individuals is more often than not mediated via some electronic platform, where transactions are often carried out remotely, and where traditional notions of trust and confidence in the identity of those with whom we are interacting can sometimes be much less reliable than hitherto we have been able to take for granted. We know that it is now easier than ever before to assume the identity of another person, sometimes with dishonest intent and, in some cases, with very serious consequences. We also accept that the lifestyle choices we like to make and societal development will inevitably increase rather than decrease the need for finding better and more flexible ways of maintaining our safety, security, and trust in the sort of everyday interpersonal transactions to which we have become accustomed.

But it is not only a question of security, for the complexity of modern lifestyles also means that questions of convenience and choice are an important part of the discussion. As we rely more and more on interactions through automated means, the challenges of coping with a rapidly growing number of different systems increase correspondingly. For example, should we try to remember a different password for each system we use, or should we reduce the chances of a lapse of memory and the consequent inconvenience by sharing passwords across systems? Are there areas of our lives where modern patterns of interaction demand greater robustness in confirming the claimed identity of the people involved—in online banking, for example? Is there some activity we now routinely carry out which was once never considered, but where security needs to be a high priority, such as protecting access to a phone or laptop? Many questions like these readily spring to mind.

All of these issues, and many other aspects of modern life, share a common feature, which is that in one way or another they all raise questions about individual *identity*—how we prove that we are who we say we are, how we can be sure that another person is who he or she claims to be, and so on. Moreover, we want to develop processes to handle these questions of identity which will work in the diverse range of scenarios in which we nowadays typically engage, which will be reliable and robust, convenient and easy to use, flexible and safe, and which do not impinge adversely on our privacy. In addition, we want these processes to be suitable for automated systems, and to operate well when those involved in a transaction are remotely located. What are the implications of these requirements, and how are we to achieve the implementation of systems which meet them? These are questions which are at the heart of this book. Let's begin our search for ways of achieving some of these desirable features by considering typical traditional ways in which we have approached these questions.

Establishing personal identity

Perhaps the most obvious way in which an individual can claim or confirm identity is through the possession of some form of *token*. Most obviously, taking a common scenario, this might be a physical key (such as the key to a house, office, car, and so on) or maybe a swipecard, and this approach is still manifestly very commonly used, at least for many routine applications. This is often referred to as determining identity by means of 'something you have', and is obviously an established, well-tested, and very successful approach, but it is also one for which weaknesses and vulnerabilities can easily be seen. For example, simply knowing that a key (or, in my own case, a smartcard programmed for an individually tailored set of access rights/permissions) has been used to effect entry to my office, does not automatically guarantee that I am the person who used it. A key can be copied, or stolen either temporarily or permanently, or I might deliberately collude with a third person for the purposes of committing a fraudulent act.

An alternative, which many people will feel has more to offer, especially for a number of present-day applications, is based on the premise of 'something you know' rather than the previous 'something you have' principle. This approach is typified by the use of a Personal Identification Number (PIN) as, for example, has been almost universally adopted with credit and debit cards, or with personalized passwords. Almost all readers will be familiar with the use of passwords to regulate access to computer systems, for example. This approach has the benefit that the 'knowledge' used is less easy to steal or copy than a physical token, and also that it can usually be explicitly chosen and personalized by the user so as to make it quite easily memorized. It also offers some choice about the degree of complexity which we wish to incorporate, maybe allowing an individual to trade degree of security provided against ease of memorization. However, it is well known that many people still write down a password, despite advice to the contrary, and it is frequently not impossible to guess what it is likely to be (many people, remarkably, still use 1234 as their four-digit PIN!), or it can be stolen through simple observation during use by the rightful owner. It is also easy to share with others as a deliberate act, and can provide very easy opportunities for collusion between individuals.

Also, in both of these approaches, an important issue which can undermine security is what might be termed the susceptibility to *plausible deniability*, which we have already touched upon. That is to say, I might be accused of carrying out some inappropriate transaction via my computer account, but my response could be to deny this, using the defence that someone else must have accessed my account without my knowledge or permission, perhaps by accidentally discovering my password or actively taking steps to find out what it is. It may not be straightforward to prove that this was not the case—and, indeed, this might actually be the truth of the situation in the end.

So we are familiar with the use of well-known mechanisms based on establishing and monitoring individual identity which are inherently, to a greater or lesser extent, fundamentally insecure. While these mechanisms continue to remain perfectly acceptable and adequate for many ongoing applications, as the extent of our reliance on automated information systems increases, and where an increasingly diverse set of applications, sometimes very sensitive, are involved, we must begin to look for more robust and reliable alternatives.

We should then ask if there is another and better alternative to the 'something you have' and 'something you know' approaches we have briefly discussed. One such alternative is a third approach based on a significantly different principle of 'something you are', and this is an approach which has led to the striking practical developments which are the focus of this book.

Figure 1 shows a (perhaps not very flattering) photograph of the author. If any of my friends or family were to be looking through this book they would immediately recognize the face in the photograph. The same would be true of the students whom I have taught over the years. The human face is, to all intents and purposes, unique to the individual who 'owns' it, and thus offers

1. Photograph of the author.

an obvious way for the individual to identify him/herself to
others—providing they are familiar with its characteristics. So,
for example, if I forget my house key and have to ring the doorbell
when I get home, I would expect my wife to let me in—because
she will recognize my face and know that it really is me. The
same would probably be true of my voice characteristics, or
indeed other observable traits. In fact, there are a lot of things
about me which are unique in this sense. Less immediately
visible to the naked eye, but offering similar opportunities for a
high level of confidence in establishing my identity, are other

characteristics such as my fingerprints, the patterning of the iris in my eye, and so on. All these features of my appearance and physiology are fundamental to my physical identity (although not always quite as accessible or as easy to check as facial characteristics), and each individually is an illustration of the idea of 'something you are', providing an opportunity to identify an individual based on characteristics which are fundamentally a part of his/her physical make-up.

This is the domain of what is now routinely referred to as *biometrics*, the technical discipline which is concerned with developing formal automated ways of reliably determining or confirming individual identity based on these personal physically based characteristics. It is because the characteristics we use in this approach are such an intrinsic and embedded feature of each individual that we should be optimistic that this third approach is, in principle, likely to prove more reliable than the earlier approaches we considered. We can easily see how this approach can overcome some of the difficulties we have already mentioned. There is now nothing to remember or forget, since our biometric data are carried around with us, while sharing my iris patterning with another person is extremely difficult unless I am explicitly colluding in some fraudulent activity, and even then it is not necessarily an easy thing to do. And deniability is now made more difficult because, if my facial image is used to gain access to a building, and the act of entering thus tightly bound to my physical characteristics, it is harder to claim that someone else had simply stolen those characteristics. Yet, as you may already have spotted, the situation is not quite as simple as might be suggested here, and indeed there are many issues which we will need to investigate further in this book.

For now, though, let us accept that an approach to individual identity based on biometrics has many potential benefits to offer.

In this chapter I will set out some basic ideas and describe a framework for more detailed exploration later.

Fundamentals of biometrics

Let's begin with some important definitions. The two most basic of these are as follows.

We may define *biometrics* in the present context as the scientific discipline which is concerned with the measurement and deployment of attributes or features of a person which can be used to identify that individual person uniquely.

Next, the term *biometric modality* refers to a particular source of the measurement data used for identification. For example, fingerprint patterns and voice characteristics represent two different biometric modalities.

There is an enormous variety of possible sources of biometric data, and thus of different biometric modalities. Box 1 lists just some of these, and it can immediately be seen that the examples

---

Box 1 Examples of the diverse range of possible biometric modalities

- Facial features
- Voice characteristics
- Fingerprints
- Palmprints
- Handwriting
- Handwritten signature
- Iris patterning
- Hand shape
- Hand vein patterns
- Keystroke dynamics

- Ear shape
- Gait patterns
- Retinal blood vessel patterns
- Odour
- ECG/EEG patterns

Most common currently:
face, fingerprint, iris, signature, voice

---

shown here, which are by no means exhaustive in their scope, cover widely differing types of data. Some of these modalities will already be familiar; others may be much less so. We will return to this issue of diversity very shortly (and at various other points in the book), but we must first consider what makes a data source suitable for use in biometrics. In other words, are there some criteria which must be met in order for a data source to be accepted for such an application?

In fact, there is quite general agreement about what these criteria should be. In principle, any human characteristic can be used as a biometric data source provided it meets the following four basic criteria (although others are also sometimes added):

UNIVERSALITY: Everyone should possess the chosen characteristic. This is because it is important that a biometric system is inclusive, and can be used by as many people as possible.

UNIQUENESS: No two individuals should be the same in terms of the chosen characteristic. If we are to identify an individual we should measure characteristics which distinguish one individual from another.

PERMANENCE: The chosen characteristic should be invariant over time. It is important that any characteristic we choose is always the same when it is measured, otherwise an individual could appear to be a different person at different times.

COLLECTABILITY: The chosen characteristic should be objectively measurable in a quantitative way. It should be defined well enough to ensure that there is no ambiguity about what is being measured.

On this basis, we can see that a number of the well-known characteristics which we commonly use in biometrics (for example, a fingerprint, an iris pattern, a facial image, etc.) all seem broadly to meet the criteria, justifying their widespread adoption in established practical biometric systems.

However, if we think about this further, we will see that we need to be a little more cautious, and perhaps that we should not interpret these criteria too literally or in an absolute sense. We can illustrate this point by considering, say, the face modality in more detail. What can we say about individual facial images in relation to these criteria?

Considering first the *universality* criterion we would probably all agree that this is generally met by this source of measurement data, in the sense that we all have a face on which, to a greater or lesser extent, various different identifying characteristics can be found. However, possessing such characteristics may not necessarily mean that they are all visible in all circumstances. For example, hairstyles might obscure certain facial features, as might wearing a hat, either as protection against the weather or as a fashion statement. Wearing sunglasses can easily obscure details of the face around the eye region, and sometimes, for religious or cultural reasons, a person may explicitly wish to cover the face.

In general terms, for the *uniqueness* criterion we would probably all say that we are unique in relation to our facial appearance, and at the most basic level this is true. However, in practical terms we may have to be somewhat more flexible. We all agree, for example, that there are often facial resemblances within families as a result of genetic inheritance. This is especially the case for twins, and if we think about the case of 'identical twins', then we quickly see that the notion of uniqueness needs to be interpreted in a rather less literal way than we might initially have hoped.

The *permanence* criterion is perhaps the most difficult characteristic of all to interpret. This is because we all know that facial appearance changes with time. We look very different in our mature youth than when we are born, and as we get older various (and sometimes quite dramatic) changes can occur. Although we expect such changes to be incremental, and probably relatively slow, they naturally have a bearing on the likely performance of a biometric system, and especially so if we aim to use such a system over a long period of time.

In relation to the *collectability* criterion for the face modality, this should be relatively straightforward, because we need only a simple and readily available transducer such as a standard camera. However, for the purposes of analysis, we may need to extract detailed information from a captured image which may be more challenging to measure (for example, 'distance between the eyes' sounds a simple enough feature, but implementing this in practice may not always be easy). Environmental conditions such as lighting may also have a very important influence on the process (look at the reflections in my glasses in Figure 1, for example) and, of course, the issues mentioned above (under the heading of 'universality') are also extremely relevant here. And we could easily extend this sort of discussion to other modalities.

Beyond these four generally agreed 'essential' criteria which must be met by a biometric modality, a number of other desirable criteria are also sometimes cited. These include, for example, *resistance to circumvention*—meaning that it should be difficult to fool a system through fraudulent attack—which is probably quite self-evident, and an intrinsic aspect of why we wish to use a biometrics-based procedure in the first place. Another is the notion of *acceptability*—that a biometric should be acceptable to the community of users for whom it is intended. While this may again seem fairly obvious, it is not necessarily clear-cut because, while in everyday situations or where a large and heterogeneous population is involved (for example, a national identity card system), it would be foolish not to take account of acceptability, there may be other applications involving targeted or specialized groups of users where a moderate degree of coercion may be exercised which runs against individual preferences (inside a prison, for example).

So we can identify a range of criteria to define a viable biometric modality, some of which may be application-specific or population-dependent, but the four initially noted above may be considered a core set of criteria which we would expect any proposed source of biometric data to satisfy.

If we look carefully again at Box 1 we quickly see that the available modalities (and we can all probably also think of others which are not included here) are very different in terms of their characteristics. For example, some (many, in fact, as we shall discover) rely on a captured *image* of some aspect of our physiology in order to determine identity (a facial image, an image of the patterning of the iris, and so on), while some use other captured time-varying signals (the voice). Some typically depend on direct *contact* with a sensor (fingerprint) while others do not (face), and so on.

It can often be very useful to categorize modalities according to these broad characteristics, since these will have important implications for how applicable a particular biometric might be in different circumstances. A contact-based biometric, for example, will limit the distance over which it can operate, since the user will need to be in close enough proximity to the sensor to make the contact required to generate the biometric sample. However, perhaps the most commonly encountered and most general categorization is to make a distinction between the following two particular types of biometric modality:

Physiological biometrics are based on the measurement of some inherent physiological characteristics of an individual. An obvious example of a physiological biometric modality is the fingerprint. This is simply part of the fundamental physiological make-up of an individual and, although it may be altered to some extent—for example by damage or injury—this is not something which is typically controlled directly by the individual.

For *behavioural biometrics*, on the other hand, the measurements involved arise from an *action* carried out by an individual, either one which is spontaneous or one which has been specifically learned. The most obvious example here is when we use the handwritten signature as a source of biometric data. Unlike the fingerprint, which is naturally and always present, the signature only exists when an individual writes it. Thus, if you were to meet me, you would not automatically be able to see my signature, or access it directly. I would need to carry out the act of writing it before it becomes available to you.

We will see why this distinction is especially useful, and why we so often try to maintain an awareness of the category into which a chosen biometric falls.

In Chapter 3 we will focus on some of these modalities in more detail, and we will see how each offers advantages and disadvantages, how each might be more or less effective in different application scenarios, and exactly how each operates. Before we do that, however, we should take a brief look at some typical application areas where biometric systems are currently used, in order to demonstrate the range and variety of applications where the topics we will study in more detail are likely to be of significant practical value.

Some applications of biometric systems

It would be easy to devote a whole book to a discussion of existing and potential applications for biometric systems, such is the variety of what is possible. By way of an introduction, however, we will briefly set out some of the current areas where biometric technologies have already become a part of our social and commercial landscape.

One of the most natural areas in which we might consider usefully deploying a technology-mediated approach to the identification of individuals is to provide a common, universal, and convenient means for citizens to demonstrate or confirm their identity in multiple different situations. National identity (ID) cards have a long history and, in some countries, have been in use for many years, in some cases dating back well before the availability of biometric technology. More recently, a number of national ID card programmes have been introduced which use biometric identity checking, the most high profile of which is probably the Aadhaar programme in India. Closely related to this area, since this too represents formal identity checking administered at the level of state governments, is in connection with travel documents, and we are increasingly seeing, for example, biometric passports being adopted, while fingerprints and facial images are commonly checked in various border control applications. In addition to the benefits accruing from an increase in the effectiveness of checking identity which a biometric system offers, this approach has also allowed a greater degree of automation to be introduced into security checks, thereby increasing passenger flow while at the same time enhancing security levels. An additional benefit with respect to travel documents such as passports is that the issuing of the original document itself is already very stringently controlled and, as we will see, the reliability of biometric systems is significantly influenced by the effectiveness of this initial step.

Beyond this, the structure of our modern society presents an almost infinite range of applications where physical access control is important. In other words, we often need to ensure that only those authorized to do so can gain entry to particular buildings, rooms, or other designated physical spaces. At one end of the spectrum of possible applications, we would be very concerned if we felt there were no checks on people entering, say, an important military facility or a nuclear power installation while, at the other extreme, we all like to protect our homes, our offices, and so on. In between these extremes there are many applications where increasingly we feel it necessary to avoid uncontrolled access, including areas within schools and hospitals, or other places where vulnerable people are to be found. Access control is vital in a number of more specialist areas too, such as protecting airside access for staff working at airports, access to police custody suites or prisons, and so on.

In a similar way, there are many applications where access to what we might call virtual spaces (where it is access to information rather than a physical location which is of primary interest) needs to be controlled. For example, there are many applications where particularly sensitive data need to be protected. Maintaining the privacy of medical data is a good example here, as is the notion of retaining restrictions on access to bank accounts and other financial transactions. At a rather lower level, many of us like to ensure that our personal computers cannot be randomly accessed by people other than ourselves and, increasingly as mobile phones (nowadays extremely versatile and very powerful information platforms) become almost universally owned, we like to be able to lock our phones.

There are many cases where accessing services needs to be limited to a specified group of individuals who are explicitly entitled to use them (entitlement to free school lunches, for example) or to public services including borrowing library books, or claiming social benefits. We can often beneficially exploit the convenience and reliability of biometric identity checking, and many applications can be found in, for example, time and attendance monitoring (for school registration, in the construction industry, and especially organizations which employ a high volume of casual labour, in the retail sector, etc.), or in the protection of personal goods or facilities (databases, memory sticks, etc.).

There is also an enormous application potential, of increasing interest in recent years, in the area of forensics. Forensic science provides a set of techniques and processes which are crucial in fighting crime, supporting the criminal justice system, and increasing security of citizens. One principal goal of forensics is to find and analyse evidence to determine the identity of individuals who have perpetrated a crime, and in this sense we can see that the study of forensics shares many of the goals of biometrics. Although until relatively recently these two disciplines have often tended to develop mostly along parallel lines, there is now a much greater recognition that the two fields have a great deal to learn from each other, and there has been a rapid growth in cooperation, collaboration, and integration of knowledge across the two disciplines, leading to increasing convergence.

Whatever the application, however, the introduction of biometrics as the underpinning identification paradigm can offer some significant benefits. These include, first, the binding of an individual to an event or data. If the fingerprint of person X is known to have been used to provide access to a high-security laboratory on a night when a security breach occurred, it is difficult to claim that it must have been person Y. Second, as we have already seen, biometrics-based verification of claimed identity should be manifestly more reliable than the more traditional techniques involving tokens or passwords. Third, as we will consider later in more detail, biometrics-based processing provides the possibility of flexibility and personal choice in a variety of identification situations.

However, reaping the undoubted potential benefits of biometrics-based systems depends entirely on ensuring that these systems are chosen, designed, and implemented in the best possible way. As with any technology, using a system inappropriately will, at best, diminish the benefits which accrue and, at worst, can be entirely counter-productive. It is this search for effective and efficient system design, implementation, and deployment which will be the focus of the remainder of this book, which will provide a detailed basic understanding of how biometric systems operate, how they can be configured and implemented to exploit their potential to the maximum, and how they can help us to think creatively about future, and perhaps as yet unconsidered, applications. In order to do this we will see that we need both an understanding of the principles of operation of biometric systems, and also a working knowledge of some key underpinning fundamental techniques.

Some context for an understanding of biometrics

We can make a number of observations about some background issues which are important in the study of biometrics. The biometrics field has a longer history than we may initially imagine, but is one in which technologies and techniques have matured very significantly in recent years. Many very successful practical systems are now being deployed, covering diverse applications including controlling access to physical and virtual spaces, protecting financial transactions, aiding forensic analysis, detecting identity theft, securing e-transactions, enhancing personal security, and contributing to the safety and protection of citizens. The explosion of electronic transactions, not just in retail and banking, but in e-health, e-ticketing, e-government, and so on, is offering increasing opportunity for biometrics-based person identification to provide robust security solutions. Research in biometrics is now also a highly internationalized and high-volume enterprise, ensuring that new applications, and better and more reliable development of existing applications, continue to emerge, and it is notable that applications range from the very small scale (for example, a small building company using biometrics for time and attendance monitoring) to major large-scale projects on a national (for example, Aadhaar, India's national ID scheme) or even international scale (for example, biometric travel documents). Also, all the predictors of the international market for the biometrics industry seem to point to a continued growth in the demand for biometrics-based products and services in the coming years.

The biometrics field is characterized especially by its interdisciplinary nature since, while focused primarily around a strong technological base, effective system design and implementation often requires a broad range of skills, encompassing also human factors, data security and database technologies, psychological and physiological awareness, and so on. Even the technology focus itself embraces diversity, since the engineering of effective biometric systems requires integration of image analysis, pattern recognition, sensor technology, database engineering, security design, and many other strands of understanding.

It is characteristics such as these which demonstrate the need for us to study the subject area of biometrics from a perspective which is wider than simply understanding a biometric system using a narrow black-box approach. We will be most successful if we can keep an eye on the bigger picture, and if we explore some key relevant, and sometimes quite diverse topics in an integrated way, which show explicitly how fundamental supporting techniques are to be used in the specific context of our chosen application area. In the remainder of the book the focus will be on the detail of biometrics—that is, the principal aim will be to provide a basic understanding of the purpose and nature of biometrics, how different modalities (i.e. sources of identification data) are used, how reliability and robustness can be assured, what limitations particular configurations impose on performance, and how biometric systems can be effectively deployed.

It is maybe worth noting too that, at a technical level, the study of biometrics is underpinned by two more fundamental areas of study, generally known as *image processing* and *pattern recognition*. Even in our preliminary survey of biometric modalities in this chapter we have already seen that many—indeed, perhaps, most—of the currently popular and widely adopted biometric modalities are based on the capture and subsequent processing of *images*. A similar study is also required of relevant aspects of the field of *pattern recognition*. This is the discipline which, as its name implies, aims to find patterns in data and, specifically in our case, will allow us to take multiple samples of data from an individual which might vary considerably because of the environment in which capture occurs, the nature of the collection process, and the inherent variability of biometric information, and then try to determine whether such variable samples may reasonably be said to be derived from the same individual. In other words, we need an understanding of basic pattern recognition techniques in order to be able to relate raw biometric data to an individual identity in a methodical and reliable way.

In concluding this chapter, let me illustrate briefly why the proposed approach is important by considering how we might check the identity of an individual through automated analysis of, say, a fingerprint image. We must first capture an image of the fingerprint of a particular individual of interest, and so we might use a sensor based on a small camera to acquire this. Now we need to compare this image against the sample(s) we have previously captured under controlled conditions (enrolment) which we know to be genuine. But maybe we find that the image we have just captured as this individual arrives for work is not very clear—perhaps his hand was particularly sweaty and greasy because it is a hot day, and so instead of seeing something like the very clear sample illustrated in Figure 2(a), we actually see a much more degraded sample, perhaps more like that shown in Figure 2(b).

It will be apparent that in these circumstances it is likely to be very difficult to match the current sample with the genuine stored sample. Indeed, there is a real problem here because it is almost impossible to see all of the detail of the fingerprint which we need for the matching process, and so we may need to apply some *image processing* to improve the appearance of the image and thereby to increase our chances of determining whether the current sample is genuine or not. Also, since we are going to

(a) (b)

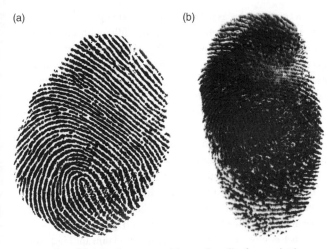

2. Examples of fingerprint capture, (a) a good quality fingerprint image, (b) example of poor image capture.

extract information from the ridge patterning on the fingerprint, then even if the image obtained is considerably better than that illustrated here, we may wish to process the image further so as to make this patterning stand out more in order to allow us more easily and reliably to find the features we are looking for.

But then there is another problem, because the nature of the capture process is such that the user will not always put his/her finger in exactly the same position, or at exactly the same angle, on the sensor capture area. Not only that, but we might imagine that on this particular day the user has a small cut or abrasion on the finger. In other words, even for genuine samples, and even when the image is of good quality, the captured sample as reflected in the image obtained will be unlikely to match exactly the stored sample which was likely to have been captured to exemplify that individual, which may well have occurred some considerable time earlier. So now we also have to cope with the issue of trying to determine how to classify a possibly wide range of different samples as belonging to the same person, and conversely deciding when a sample is too different from previously acquired sample(s) to allow us to say it is genuine. This means we need to develop some techniques for grouping samples together as belonging to the same or different *classes*, and it is this that takes us into the study of *pattern recognition*, which offers us robust techniques for tackling exactly this problem.

So, this book will take us on a journey, leading us through both fundamental principles and also some specific application-related studies, to an understanding of how biometric systems work and allowing us to judge for ourselves whether benefits are to be found in adopting such systems and what these benefits might be. We will also then be able to take a look into the future, and know better how to assess and evaluate the claims for biometric technology which we are increasingly reading about in the newspapers and seeing in film and on the television. In Chapter 2 we will delve more deeply into the ideas introduced here, so as to understand the structure of a biometric system, how it operates, and how we can begin to describe and evaluate its performance.

Chapter 2
Biometrics: where should I start?

Introduction

We have seen in Chapter 1 that the field of biometrics is concerned with the measurement of personal physical or behavioural characteristics which we can use in order to identify an individual. We have seen how this should in principle be fundamentally more reliable and robust than the more traditional ways in which we have tried to determine or confirm individual identity. However, taking this approach will mean that we need to develop a new way of thinking about the identification process, and that we will need to design systems to achieve our goal which are rather different from those we have considered in the past.

In this chapter we will begin the process of understanding what constitutes a practical biometric system. We will look at the principles on which such a system will operate, build up a picture of the components we need to construct such a system, and take the first steps towards understanding how to implement a biometric system. We will also try to understand why system performance may not always be error free, and we will look at how, when, and why errors can arise. In order to put these issues into some sort of perspective, we will also develop tools which will allow us to evaluate, both qualitatively and quantitatively, the performance of a biometric system, to understand more clearly the nature of the interaction between the user and the system itself, and to determine helpful ways of describing both basic system factors and user characteristics which will ultimately influence system performance.

Biometric system basics

We first need to be aware that in implementing a biometric system for person identification it is necessary to distinguish two distinct *operational phases*, which we can characterize in the following way.

First let's consider what we might call Phase 1 of system operation. In order for us to be able to identify an individual based on biometric measurements, it is obvious that our system needs to have some idea of what measurements are to be expected from any given individual entitled to use the system. We need to establish this by asking a potential user to provide samples of biometric data, derived from his/her physiological or behavioural characteristics of interest, for use in defining that individual in terms of the measurements we propose to use.

For example, if we use the fingerprint modality we will need to ask the potential user to give us at least one sample of his/her fingerprint, against which we can compare future samples presented by a questioned user. In fact, rather than just store this single specific sample, we may prefer to collect several samples (which will allow us to capture information about the sort of variability we can expect to occur when samples are donated at different times, when interaction with the sensor may vary, or where environmental conditions may change). We then need to construct a *model* of that user with respect to the measurements obtained, which captures the variability we can expect naturally to occur. This model is often referred to as a *reference model* or *template* of the data which characterizes that user. This reference could simply consist of a set of the different donated samples, or could consist of a different kind of model which might be, for example, a statistical model which encompasses in a more mathematical way the variability across several samples.

This process of collecting genuine user samples and setting up a template for a proposed system user is generally referred to as the *enrolment* process, and thus this phase of operation, which is a prerequisite for ongoing use, is the *enrolment phase*. It is worth noting that this phase needs to be carefully controlled if, in subsequent use, the biometric system is to be used reliably for identity checking. This suggests two principal issues which we need to keep in mind. First, particularly if we are dealing with a high-security application domain, we need to have in place a robust procedure to ensure that, before biometric data are extracted and used to construct the reference template, we can be sure that the enrolee has had his/her identity carefully checked in an alternative and appropriately rigorous way. This is a little like the procedure currently used to obtain a physical passport—it requires a careful and intensive procedure on application, but thereafter the passport itself can generally be used easily and with confidence. And, of course, the effort expended at this point can be determined in relation to the importance and sensitivity of the intended application. The second issue is that we have to accept as a consequence that enrolment may be time-consuming, but we must control the nature and quality of the data captured, simply because subsequent system reliability and performance will be severely compromised if a poor enrolment is achieved. And, as noted, enrolment is generally a one-off operation, after which we can expect using the system to be maximally reliable.

When Phase 1 is completed we have a system which 'knows' each individual who has enrolled (in the sense of having stored genuine biometric data about each one), and which is therefore capable of being used in a particular application environment as a means of checking the identity of future users. Specifically, it will be able to recognize those individuals who have already enrolled. We can then turn to the second phase, which concerns the operation of the system when in use as a means of ongoing identity checking.

Subsequently, then, in what we might refer to as Phase 2 of system operation, the system into which a number of entitled users have now enrolled can be used to check the identity of those who use it. Hence, when a biometric sample is presented to the system, we need to invoke a procedure to determine either

(a) whether or not a sample presented corresponds to one of the reference models already stored within the system, one model for each enrolee

or

(b) the likely authenticity of a presented sample in relation to a claimed identity. This requires us to check a sample presented against just one of the stored enrolment templates.

In either case, we will need to carry out a *matching* (*comparison* is now more formally used) operation to compare the current sample with reference templates constructed during enrolment.

We will return to these issues in the rest of this chapter, but we must first examine further what the structure of a biometric system might look like.

Structure and components of a biometric system

Figure 3 shows a schematic representation of the major components of what we might expect a biometric system to consist of. We can identify some general blocks in this diagram which will demonstrate the various key stages of data processing which are likely to be required, and which will help us to understand the techniques and processes examined in more detail later. In Figure 3, the input to the overall system is the presentation of a source of a biometric measurement (for example, this could be the presentation of a fingertip at an appropriate capture point). The

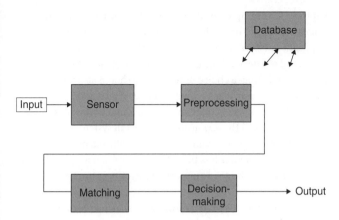

3. Basic components of a biometric system.

overall system output is the result of a process which compares information extracted from the current input sample with information collected at enrolment. This may take a variety of forms, which we will consider further once we understand more about the overall processing chain.

Let's consider briefly each of these system components in turn.

The *sensor* provides the link between the physiological site from which the proposed biometric information is taken (or the behavioural activity which generates appropriate data) and the processing function which assesses the data collected. It thus captures 'raw' data and passes this into the processing chain. The sensor is often a camera (for the face or iris modalities, for example), but could alternatively be a microphone (for the voice modality), a standard off-the-shelf peripheral device (e.g. a standard graphics capture tablet for handwriting capture), or even a special-purpose device constructed explicitly for a particular application. In fact, it is apparent that many different possible sensors may be available, even for a single modality (a variety of technologies can be deployed in fingerprint image

25

capture, for example), and the characteristics of a particular sensor will be important in determining its suitability for a given application. Issues such as the quality of the data generated, the size of the device, its cost, and so on, must all be taken into account in making a choice of sensor. We should also remember that the sensor provides the precise point of interaction between user and system, and is therefore fundamental to the question of what we often call the *usability* of the overall system. In other words, if the interaction with the system is difficult, or inconvenient for users, the system is less likely to be used as a matter of choice, so this could also be a very important factor to consider.

Let's now consider the *preprocessing* stage shown in Figure 3. The data which are acquired at the sensor are often not in an ideal form to guarantee the best and most accurate matching operation which is required at the heart of the biometric system. The data may have been collected in poor environmental conditions, perhaps there is some noise contamination, or maybe the contrast of the image acquired is poor, impairing the visibility of important characteristics.

Whatever the cause, we may need to carry out some appropriate 'preprocessing' initially on the raw data in order to counteract the degradation caused by these different effects. If we take as an example the fingerprint modality, we can implement algorithms to achieve useful modifications to the content of the captured image (for example, thinning the lines in the image so that the fingerprint structure can better be seen, removing noise which is contaminating the image and causing misleading detection of the features of interest, or enhancing the contrast of the image, the better to reveal the patterning which we are trying to analyse). We should also note that the exact preprocessing operations adopted will depend on the modality of interest, the sensor used, the environmental conditions, and so on.

Although we could actually represent this as a totally separate component, we will include here the fact that in most biometric processing we will not be using the captured information (image) directly, but will instead extract from the image a set of specific measurements (*features*) which we use to characterize each individual user. If we are considering the face modality, for example, we might extract a measurement of, say, the distance between the eyes as one component in a list of features to describe each individual face. In Chapter 3, which deals with specific modalities, we will learn more about what sort of features are appropriate in each modality.

At the output of this part of the processing chain we will have generated a set of measurements (optimized as far as possible, we hope) which we can use to form a model of a user-specific fingerprint pattern (if this is the adopted modality) and which is stored as a template, characterizing a particular user of the system. The next step in the overall processing chain requires a *matching* (*comparison*) operation. Having acquired from the system user the information which we are to use to characterize that individual, the next step is to try to match this to one of the reference models (templates) constructed during the enrolment process for all those who are entitled to use the system.

In the matching stage, therefore, we will examine the data acquired from the input for the current system user and compare this against the reference model(s) appropriate for the task in hand. The basis of this comparison is to discover whether the current input data provides a match against one of the templates currently stored. If we do find a match, then we can say that the current user is most likely to be the individual whose template was matched by the input data. In other words, the aim of the matching operation is to determine whether the test (input) and reference (enrolment) data can be considered as belonging to the same person. This leads us to another very important and fundamental issue, since it is very unlikely that we will find an absolutely exact match, for the reasons we started to outline earlier (lack of reproducibility in data capture conditions, changing environmental factors, natural changes in the physiological or behavioural traits which define the chosen biometric modality, and so on). This is a key question which we will revisit later in this chapter. For the moment, however, let us assume that the output of this stage will be information about how well the current input data match an appropriate template or set of templates.

We then come to the *decision-making* stage. Depending on the outcome of the processing at the preceding stage, various possible decisions may be generated. In most cases, however, this will be influenced by a somewhat more subtle factor, which is the *degree* to which a match can be said to have been found. This idea will be clearer if we consider just some of the possible outcomes which can be regarded as *decisions* appearing at the system output. These possible outcomes may include, for example, the identification of the user providing the test sample at the input, a confirmation or rejection of a claimed identity, or maybe a measure only of the likelihood that the current user is a particular enrolled individual. Also, a further possible outcome, especially if a clear-cut decision is proving difficult, might simply be a request for the user to provide a further (and better, we hope) sample. Other possibilities can also be considered, and so we see that the exact form of the system output is not defined uniquely for all time, but can be chosen to meet the requirements of a particular application.

The final element shown in Figure 3 is a *database*. It should be self-evident by now that the system will require an appropriate database, since this will be needed to store the relevant biometric data provided by all those individuals who have enrolled on the system. This will interact particularly with the *matching* and *decision-making* operations already described.

The database will accept new enrolees as determined by the system operator/owner, and may also include other information (demographic or other relevant details) in addition to the purely biometric data. The database could be an integrated part of the biometric system locally, or could be held at a remote location and accessed as required by the system. However, in whatever way it is configured, the security of the stored data is paramount to protect all those enrolled on the system, and in Chapter 4 we will look more carefully at some of the implications of this requirement, which is vital to the trust and confidence which users have in the overall system.

It is worth emphasizing again that a satisfactory enrolment is fundamental to good operational performance of the overall system. In other words, especially in generating and acquiring the enrolment data which the database will hold, a rigorous procedure must be in place. This will both provide confidence in the credentials of the enrolee in the first place, and will also ensure that the data collected are of the highest quality possible. 'Garbage in, garbage out' is one of those clichéd phrases which really does apply here.

Before we leave this section, we can now develop our thinking a little further, even before we move on from the basics. It will be apparent already that there are (at least) two different possible scenarios to consider. To use the conventional terminology, we may wish to execute one of two possible functional operations using the biometric system, which we can easily explain as follows.

The first of these corresponds to an *identification* process. In an identification scenario the aim is to answer the question 'who is this?' in response to the presentation of a sample generated by a person of unknown identity. It involves the comparison of the current sample acquired with all of the stored reference models in the database, to determine which (if any) template is matched to the current sample (or, at the least, which template is most closely matched). The identity of the owner of the sample is thus determined to be that corresponding to the person whose reference template is matched (or most closely matched) in this comparison process. The user is *identified* by means of this process.

Alternatively, we can consider what we call a *verification* process. In this alternative scenario, the question being asked when a sample is presented to the system is slightly different, and can best be stated as 'Is this person who (s)he claims to be?' This involves the user in claiming a specific identity ('I am Person X'), at which point the current sample must be compared with the reference template provided at enrolment by the person whose identity is claimed (Person X in this case). If a suitable 'match' is found between these two then the claimed identity is confirmed. The claimed user identity is *verified* by means of this process or, alternatively, is refuted if no appropriate match is found.

Intuitively, we would probably agree that reliable operation for the second scenario is, in principle, easier to accomplish than for the first. However, both scenarios provide challenges, as we will discover.

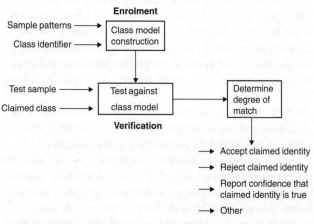

4. Schematic representation, at the operational level, of a verification scenario.

To conclude this section, Figure 4 shows another schematic, which illustrates the procedures we have been describing above, framed in terms of a set of low-level operations, and we have focused here on the verification configuration.

We are now in a position to develop in more detail our understanding of the issues involved in working with this processing chain in practice.

Practical and operational considerations

We will begin by looking more carefully at what is involved in the matching procedure we have introduced in this chapter. As we have noted, we generally are not looking to find an absolute or exact match between even a genuine input sample and its corresponding template, for all the reasons mentioned. Indeed, if we do find an exact match then we should probably be a little suspicious, because such is the inevitable variability of individual samples, environmental conditions, and so on, that such an occurrence can be considered very unlikely. Instead, we are more interested in looking at the *degree of match* between the input sample presented (questioned sample) and the template against which it is being compared. In fact, in the end, what we will be mainly concerned about are the statistical distributions of the match measures generated over a range of both genuine and non-genuine samples.

Hence, we will aim to compute a matching *score*, representing the degree of match found between input and reference samples. This can easily be, and usually is, seen as a measure of the *similarity* between input and reference, but could equally be represented as a measure of their *dissimilarity*. To produce this matching score we will need to define an appropriate distance metric, which will depend on the nature of the features extracted and data stored (and is therefore to some extent modality dependent). Then a 'good' match will be characterized by a high similarity score (or, conversely, a low dissimilarity score), and vice versa in the case of a 'poor' match.

However, the main point here is that the question of whether the match is or is not good enough to confirm identity is not an absolute issue, but is determined by the operator of the system, who must set a *threshold* of acceptable similarity for the purposes of making such a decision. This is an important idea, because this threshold can be chosen to reflect the nature of the application for which the system is being used. In high-security applications, for example, this threshold should be set pretty high (a very high degree of match should be required before identity is confirmed) while in low-security or non-critical applications, the threshold can be set much lower, with practical implications which can easily be anticipated, and which we will come to.

If the system is configured for the task of the identification of individuals (i.e. we are comparing the input sample against the templates of all enrolees) then we could simply choose the template which generates the highest match score as representing the identity of the input sample, but we will nevertheless probably also wish to set some minimum threshold score which must be attained, just to ensure that we are not accepting a dangerously poorly matching sample, which is very likely not actually to belong to any of the enrolled individuals.

We can see, then, that the notion of two samples matching or not matching is not entirely straightforward, and needs to be managed in the context of each particular system and application. Beyond this, we also need to recognize that errors will sometimes occur, and it is important that we explore this further too, since understanding error scenarios is important to the practical deployment of a biometric system. So let's examine this issue for illustrative purposes in the context of a verification scenario. In fact, there are two main types of error which we can most usefully look at initially. We can describe these in a general way as either Type I or Type II errors, explained as follows.

A Type I error is an error which occurs when a genuine user is judged by the system to be an impostor, and the claimed identity of the current (genuine) system user fails to be verified according to the match score generated. Here, if, through extensive testing, we are able to measure the proportion of verification events for which this occurs, then we can compute what is in this context commonly referred to as the *False Rejection Rate (FRR)* of the system.

The second common error characterization is a Type II error. A Type 2 error occurs when an impostor is judged by the system actually to be the individual whose identity has been (falsely) claimed. This impostor is then therefore incorrectly accepted as a genuine user. This leads to an error rate measure commonly referred to in this context as the *False Acceptance Rate (FAR)*.

In fact, we can illustrate the typical effect of these different error types as shown (not very scientifically!) in Figure 5.

(a) (b)

5. The possible effects of Type I (FRR) and Type II (FAR) errors?
(a) Type 1 error scenario, (b) Type 2 error scenario.

Just to dig a little deeper at this point, in this sort of situation we may find it beneficial to distinguish between a fraudulently produced sample generating, say, a false acceptance, and one where the error-generating sample is accidentally wrongly matched, but where no fraud was intended. This first case is relatively straightforward and, if we use the handwritten signature modality to illustrate, corresponds to one writer aiming to forge the signature of another in order to gain unauthorized acceptance by the system. We might refer to this as an *active forgery*.

The second case might arise for a number of reasons, which are especially easy to understand in the case of this (handwritten signature) modality.

For instance, for many people, the signature they adopt is just a literal writing out of their name. However, names are not unique and can be shared across many different individuals, although obviously most often within cultural or geographical groups. In the UK, for instance, one of the most common family names is 'Smith'. Likewise, one of the most common given names is 'John', and thus we can safely assume that multiple individuals with the name 'John Smith' are likely to be found within any large population of UK system users. Unless they have very different writing styles, or unless we are very careful in our choice of features on which to base an analysis of individual writing, therefore, it is almost inevitable that someone signing themselves 'John Smith' will at some point produce something rather similar to another completely different person of the same name who also signs his name as a written concatenation of the letter strings 'John' and 'Smith'. The problem is exacerbated in this example also because the two components of this name are particularly short, offering minimal scope for any small execution differences to be emphasized. We can thus clearly see the opportunity for one writer's signature to match another's unintentionally, and certainly without any fraudulent motive. Thus, although such a sample might be seen as a forgery, we would prefer to refer to this as a

6. Illustration of a potential problem in handwritten signature processing.

passive forgery. Figure 6 illustrates this issue, showing examples of the signatures of different writers with the name 'John Smith'.

These examples are among the most striking illustrations of where we might need to distinguish between different types of 'impostor', although similar scenarios can be found in other modalities as well.

Returning to the main discussion, in fact we can easily experimentally measure quantitatively the error rates generated in a biometric system provided we have available a large enough set of samples of known identity on which to conduct an experimental study. Furthermore, we can see that it is possible to engineer a trade-off between these two types of error, and that we can achieve this simply by the choice of the threshold we set in determining the acceptability of the degree of match between test sample and reference sample. In other words, we just have to decide what level of similarity is acceptable to confirm claimed identity (or make an identification) in a particular application. This is what will determine the two different error rates associated with a system.

We can also see, as has been mentioned, that exactly how we make this trade-off, and thus where we set the acceptability threshold, will depend on the application scenario of interest. For a high-security application users would probably be prepared to accept the frustration and inconvenience of a higher probability that they are (incorrectly) not accepted by the system in return for the knowledge that it will be very unlikely for an impostor to be accepted, while in more routine, lower-security situations the opposite prioritization would more likely be required. In these latter circumstances the benefits of ensuring that no unauthorized user is accepted might be vastly outweighed by the inconvenience caused to authorized users who are incorrectly not accepted on frequent occasions.

In fact we can look at these relationships graphically to generate an easily assimilated system performance profile by plotting the measured error rate of interest (FAR or FRR) as a function of the choice of threshold, which we will here designate θ. Figure 7 illustrates the way in which FAR, for example, would typically be expected to vary as a function of θ. We can see here that if we demand only a low degree of similarity between input sample

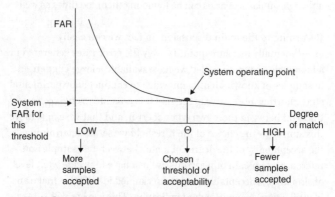

7. **Typical variation of FAR with chosen threshold. Here the FAR (y-axis) at threshold θ (x- axis) is marked.**

and reference template for acceptance, the potential for an impostor to provide a sample which matches according to this setting can be relatively high, leading to a high measured overall value for FAR. On the other hand, if we raise significantly the degree of similarity required for the system to accept the claimed identity (by increasing the value of θ), then it becomes progressively less likely that an impostor will be able to produce a sample of sufficient similarity to exceed the threshold. The FAR then correspondingly decreases.

In Figure 8 the typical curve for FRR as a function of θ has been added to the FAR curve shown in Figure 7. We can see that in the case of FRR the situation is reversed compared with FAR. If we set the acceptable match threshold low, then even poor samples generated by the genuine user will easily be accepted, meaning that it is very unlikely that the genuine user will not be accepted by the system. This results in a low FRR for small values of θ. However, conversely, if we raise the threshold value, thereby requiring a much higher degree of similarity of the input sample to the claimed template in order to define a match (i.e. the genuine user needs to generate a more accurate reproduction of the reference enrolment samples), then (s)he may find that the samples presented, even though genuine, do not reach the threshold of acceptability. The FRR will therefore rise accordingly.

There are two particularly important features to note about this graph, as follows. First, unsurprisingly, we see that FAR and FRR change in opposite directions with θ. This shows exactly, for any given system, the quantitative specification of the trade-off which can be achieved between these two error rates.

Second, the point at which the FAR and FRR curves cross over is marked on Figure 8. If we read off the actual value of the error rate at this point, which is where the FAR and FRR are equal, then we have a value for what is generally referred to as the *Equal*

Taking FAR <u>and</u> FRR

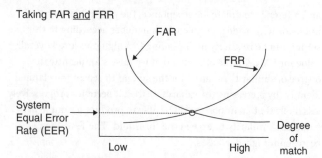

8. Variation of both FAR and FRR with chosen threshold.

Error Rate (EER) for the system. This is a useful and frequently adopted performance metric for a biometric system.

Also, it is helpful to note that although the terminology introduced in the preceding paragraphs is still quite common in the many available books and articles which discuss operational biometrics, increasingly, slightly different and perhaps rather clearer terms are becoming adopted for the two principal error rates we have specified here as FRR and FAR. So we are also likely to encounter the term False Match Rate (FMR) in place of False Acceptance Rate (FAR) and False Non-Match rate (FNMR) in place of False Reject Rate (FRR).

There are different ways of depicting the performance curves in Figure 8. For example, we can instead use FAR (FMR) and FRR (FNMR) as the axes of the performance graph, and mark the corresponding FAR(FMR)/FRR(FNMR) values measured at different chosen threshold values. This will produce an alternative graphical representation of error-rate performance such as that shown in Figure 9, which is generally referred to as the Receiver Operating Characteristic (ROC) curve. In fact, depending on the exact way in which the axes are labelled, we will sometimes find this sort of method of representation of the error category relationship referred to as the Detection Error Trade-off (DET) curve.

We could plot FAR against FRR for differing values of match threshold (θ), as follows:

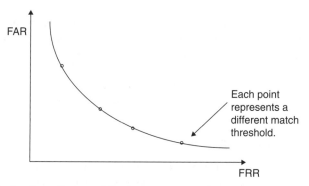

9. **An alternative way of showing error-rate performance.**

There are some other useful error measures which might be helpful in explaining how a biometric system performs. While the fundamental and most common FMR/FNMR measures already mentioned represent the most common metrics adopted to describe biometric system error-rate performance, a number of other measures can also be very useful in capturing relevant information about system performance.

Two common examples are the *failure to acquire rate* and *failure to enrol rate*. The failure to acquire rate represents the proportion of times the system fails to capture a sample when a user attempts to donate one. This might be caused by a range of possible malfunctions, such as a problem at the sensor, a quality check failure, and various other possible issues. The failure to enrol rate is the proportion of potential users who cannot be successfully enrolled (maybe because of interface issues, data quality again, etc.)

So we can readily see that a range of simple metrics can be derived for any given system, based on the availability of a database which contains labelled samples, or by capturing some record of the overall 'history' of a particular system.

Collectively, such metrics can help us to build up a picture of the operational profile of a system, and therefore be in a good position to determine whether a specific system will be able to meet a set of specifications required by a given application. Indeed, such metrics allow us to specify the requirements of a system design from an analysis of a particular task domain in the first place.

Typical sources of error

We have seen that a biometric system may not give us perfect performance, and so we must try to understand the nature of error sources as much as possible, take these into account in specific applications, and hence try to find ways of managing a system to handle effectively the impact of any errors which do occur. It is worth investigating these issues further to see if there are any broad guidelines on which we can build application-specific solutions.

Let's first consider the obvious question of exactly why errors arise in a practical situation. If we were using one of the more traditional approaches to identifying individuals—a traditional PIN-based or password-based system, for example—then we would expect to find that analysing the decisions made is a much more black-and-white situation. Thus, a password provided by a system user either matches exactly, or does not match at all, that which is stored as a reference. But, as we have seen, this is not the case with biometric identity checking. Biometric data are generated by the measurement of physical (or behavioural) characteristics of an individual, but this approach carries with it a number of possible uncertainties.

Obviously, the sensor adopted for data capture may introduce distortions of the data (e.g. camera resolution), while short-term changes (through accident or illness, for example) at the source of the physical data may temporarily change the specifics of the data acquired. Longer-term changes (e.g. the natural ageing process) may cause ongoing variability in the acquired measurements, and changing ambient/environmental conditions (lighting, temperature, noise, etc.) can cause significant perturbations in the data collected, depending on the sensor used.

At a rather different level, problems of interaction between user and system can introduce data quality compromises. This can arise from purely physical causes (for example, the layout of the data acquisition system may not be conducive to normal intended use) or for reasons which are more at the user cognitive level (for example, the instructions to be carried out by the user are not clearly communicated, or fail to be understood by the system user). We may find that in designing our system we have used inappropriate test data, or maybe simply have available insufficient data to ensure we are able to capture an understanding of the actual variability of acquired samples in normal use. In a similar way, we may have in the user population some individuals who just have unavoidable naturally highly variable biometrics. This is most commonly experienced (though not exclusively so) and likely to be most marked in behavioural modalities (for example, a user may not have developed a consistent and repeatable form of her/his signature, or maybe someone has, for whatever reason, changed the signature form used).

In other words, the fundamental problem we face here is that biometric data are *inherently variable* in nature. This is why, when it comes to biometrics-based identity checking, we have to move away from the old notion of absolute matching of input sample and stored reference, and instead adopt the idea of a matching score to check against a threshold of acceptability chosen to suit the requirements of a particular application.

In order to illustrate these ideas, let us consider typical sources of data variability in two common biometric modalities.

Likely sources of data variability for the face modality

If we are working with the face modality we can easily identify a number of sources of likely data variability. Some obvious examples include the following.

The pose of the user to the camera (full face, looking left or right, head tilted up or down, distance from camera, etc.) may vary, presenting a range of different images to the camera for the same user. Similarly, we can see that the facial expression of the user will modify the appearance of the face, and may distort some measurements which otherwise we might expect to be characteristic of an individual, while the colouring or texture of the face may change in some circumstances (for example, after excessive exposure to the sun).

The nature of the image of a face which is captured can be rather sensitive to the ambient lighting conditions under which the facial image is captured, and this may change, either through natural cycles of light and dark, increasing or decreasing amounts of sunlight incident in different weather conditions, and so on, or because artificial lighting levels change with time (day-time vs night-time, for instance).

Cultural and stylistic conventions may also be a factor. An obvious example is that a user's hairstyle may be changed, affecting overall facial appearance, or may sometimes obscure features which are important in the matching process. Likewise, whether we are or are not wearing spectacles will change, sometimes quite markedly, the appearance of the face or, again, may interfere with the process of acquiring the required measurements.

42

Likely sources of data variability
for the voice modality

Similarly, it is not difficult to identify some examples of likely
sources of data variability when using the voice modality.

First, we can easily appreciate that the sound of our voice
can naturally change over the course of a day, often quite
significantly—as we become tired, for example, or—very familiar
to the author—after an extensive spell of talking (giving a large
number of student lectures springs to mind!). Alternatively, some
readers will be aware that the knock-on effects of, say, a hangover
after over consumption of alcohol, can significantly modify vocal
sound. In a rather similar way, a sore throat or a head cold often
change the functional structure of the oral and nasal cavities
(nasal congestion, for example) and will therefore change the
nature and quality of the vocal sound produced. Perhaps less
immediately obvious is that our voice characteristics can be
affected by changes in our mood (for example, anger or irritation)
or psychological state (for example, if the speaker is nervous or
under pressure).

In terms of perhaps more controllable factors, if we are operating
in a noisy environment then this will introduce further distortions
of the measurements we need to make. And, both here and in
the example of working with the face modality, the quality and
positioning of the sensor adopted will have an impact on the
nature and quality of the measurement data generated, so the
set-up of the physical environment in which the system will be
operating always needs careful consideration.

Other relevant operational factors

There are other factors also to keep in mind when working with
a biometric system. First, in designing a biometric system we

ultimately have control over exactly what characteristics or *features* we extract from the captured raw data, and understanding the source of possible errors will help us in making a wise choice of which features we should aim to work with.

Sometimes we will be able to control some of these factors. Considering another example, we know that in making an application for a conventional passport we are required to comply with predetermined specifications about pose, expression, and so on in relation to the photograph we submit, and these principles can sometimes be extended to scenarios involving automated biometric systems. However, this may not always be possible and, in any case, some of the error sources referred to in the preceding sections will still be largely out of our control.

Nevertheless, we can learn some important lessons by considering these issues. This is because when we design a new system we can make choices about what sort of data we feel it best to use and, most importantly, as we have noted, we can choose the features which we propose to measure and use in the identification process. This choice needs to recognize that different features (measurements) may be more effective or less effective in different applications and in different modalities. Specifically, we need to think about two factors which will inevitably affect the performance of any biometric system, but which arise principally from characteristics of the particular set of users who enrol. In this respect, we need to consider two primary areas of variability in relation to biometric data.

First, we need to think about *intra-user variability* with respect to biometric samples. Different samples from the same individual will (unavoidably) each be different to some extent, and if these differences are large, we cannot be sure whether we are looking at natural differences in the same person or, in fact, differences occurring because we are looking at samples from different people. We should additionally be aware of the nature of *inter-user variability* in biometric data. Samples from different users will also be different, but that can be seen to be a good thing, since we want to be able to distinguish between one individual and another based on what we measure.

This leads us to a particularly important guideline for system design: a 'good' biometric feature will, ideally, show LOW intra-user variability (we would ideally like samples taken from any individual at different times to be as similar as possible, thus tightly defining the expected characteristics of that individual) and HIGH inter-user variability (we would like different individuals to appear as different as possible with respect to their biometric measurements adopted, to make it as easy as possible to distinguish between them). This is a very useful general principle which can guide us in deciding on what biometric features to use for any modality and for any application. And, as already mentioned, it is probably wise to be suspicious of an exact match between an input sample and a stored reference sample. We will look at some typical features adopted in relation to different modalities in Chapter 3.

User population characteristics

In most applications of biometrics, it is likely that we will have to deal with a population of users which is typically highly non-homogeneous. This non-homogeneity will be related not only to obvious factors such as gender, age, ethnic background, and so on, but also to differences in the nature, quality, and variability of the data which individuals generate. Hence, the 'biometric performance', as it were, of individuals within a given user population can usually be expected to vary considerably. A handy way of describing key sub-groupings within an overall population involves using animal names for categories with properties suggested by those animals. The four most encountered categories are usually defined as *sheep* (most common user type, generally perform well, and are broadly problem-free), *goats* (biometric measurements tend to be very variable, typically generate large intra-class variations), *lambs* (relatively easy to imitate), and *wolves* (can easily appear to be other users, present a greater threat of attack).

So when we are considering the performance of a system, it's useful to bear in mind that the constitution of a given user population can influence significantly the difficulties which might be encountered in use in terms of the balance between types of error. Similarly, it can be very helpful to know to which category in this so-called 'biometrics menagerie' an individual system user may be considered to belong, allowing us better to understand the way the system is likely to perform for that person, and ensuring that, where necessary, we can make appropriate arrangements to optimize the way in which that user can benefit from the system features provided.

Considerations such as these move our discussion on a good deal from the basic principles of biometrics established in Chapter 1, allowing us to drill down further into how a biometric system functions and what sort of components it consists of. A good understanding of the basic structure of an operational system is essential in providing an opportunity to specify a number of very useful metrics which can be used to describe system performance and therefore—importantly—which we can now also use to *evaluate* a particular system. This is crucial in determining whether or not a given system is likely to be suitable for an intended application, but also gives us some tools to use when we set about actually trying to specify the performance criteria which we expect a system to satisfy in a particular application.

So far, we have tried to explore operational principles of biometric systems in the most general way possible, largely independent of any one modality. Let's now move on to look more closely at a selection of specific modalities, to understand something of what is involved in adopting them, and to explore what particular factors are most important when we choose any particular option. We will begin by taking a look at some of those most commonly used in current applications.

Chapter 3
Making biometrics work

Biometric modalities

Now that we know more about the most general operational principles for biometrics-based person recognition, we can explore in greater detail the various options which we can consider in choosing a modality. In other words, what particular individual characteristic should we measure, and what does this entail?

We should bear in mind that the different modalities each have different properties, providing a better or worse fit to a particular operational domain and set of application requirements. We have seen some of these properties in Chapters 1 and 2: for example, some modalities may be considered *physiological*, relating as they do to the direct measurement of a characteristic which is embedded in the physiological make-up of a given individual. Other modalities can be considered to be *behavioural*, because the data they generate can only become available as a result of some activity carried out by an individual. Such a distinction can be important in considering which modality to adopt, often for very basic reasons. Collecting a signature (behavioural) will usually take more time (because the execution might require several seconds as the writer forms the signature through a series of hand/finger movements), while a fingerprint (physiological) can be accessed merely by touching or swiping a sensor in a single one-off movement.

Other factors might be important too, including how simple or complex is the technology required for accessing the required data. Capturing a facial image requires only a camera, a ubiquitous, commonplace, and well-known capture device (which is also quite general-purpose and therefore relatively cheap), while capturing a signature may require a special-purpose tablet to track pen movements during execution (although, of course, a simple camera can also capture the completed signature—with implications which we will discover shortly). The progress of technology as a result of rapidly developing mobile information platforms such as phones and computer tablets has, however, resulted in signature capture becoming progressively more simple and widespread.

Signature capture in its most usable form also requires proximity to the capture sensor, and preferably a stable surface on which to write, if we want to avoid potential distortion of the data generated (although signature acquisition from gestures tracked entirely in the air is also now a possibility), while facial image capture can be achieved at greater distances between individual and sensor. Fingerprint acquisition is generally overt (though it is not impossible to contrive a situation where a subject is encouraged to provide a fingerprint without explicitly being aware of this), while it is simple to capture the gait patterns of a walker covertly. Indeed, all the possible modalities have a number of advantages and disadvantages which need to be considered (and sometimes traded off against each other) in making a practical choice. And, unsurprisingly, cost will always be a factor, generally being less of an issue when widely available, general-purpose, and cheap technology can be used (e.g. a camera) than when something more specific or special-purpose is required.

In order to assess the practicality or suitability of a particular modality in any given application, however, or to judge how effective biometrics-based identity monitoring is likely to be in a particular situation, we need to understand something about the major issues which affect a chosen modality, and thus it is useful to understand more about how each works. It is not possible, in this short introduction, to investigate all available modalities, or even to do more than outline some basic principles. Nevertheless, it is possible to understand the essentials, so in this chapter we will select four major modalities and examine more closely their relevant technologies, to provide an understanding of the fundamentals of operation within the more general framework introduced in Chapter 1 and Chapter 2. These specific modalities are chosen, first, to ensure that we have covered some of the most well-established modalities currently in use and, second, to illustrate the variety of modality types available.

Fingerprints and fingerprint processing for biometrics

One of the great advantages of the fingerprint modality is that the concept of using fingerprints to recognize individuals is well known to most people and, moreover, we know that fingerprint checking has a long and successful history (more than a century, in fact). Indeed, long before fingerprints became such an important part of the biometrics landscape with the advent of automated fingerprint processing, human fingerprint experts were an important part of identity-related scenarios, most commonly as part of criminal investigations. In fact, this was also to some extent an early disadvantage in the development of automated biometrics, since there was some initial resistance to the use of fingerprints (strongly associated with crime and with strong policing connotations) in the often more innocent and everyday activities targeted by biometrics. With the passage of time, however, this has become much less of an issue, and fingerprint-based biometric processing is now widely accepted and adopted in a variety of applications, not least, but not now limited to, those within the criminal justice system.

The adoption of the fingerprint as the basis of establishing identity, as with all biometrics-based information sources, assumes that

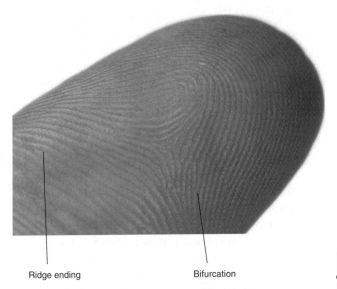

Ridge ending   Bifurcation

10. Examples of minutiae marked on a real fingerprint.

the fingerprint is, to all intents and purposes, unique to an individual. Figure 10 shows a typical fingerprint which, as can be seen, visually appears as a set of so-called *ridges* (the visible lines) and *valleys*, the spaces between the ridges. Less visible to the naked eye, there are also lower-level characteristics, such as sweat pores and other features, which are less easy to capture as they are smaller and may be largely sub-cutaneous, and therefore require more specialized acquisition techniques, but we will focus here only on the immediately easily visible ridge patterns.

The first step in the practical deployment of the fingerprint in biometrics is to acquire the details of the ridge pattern. There are a variety of ways in which this can be done, but it is obviously possible to use a camera system for this purpose, acquiring a simple visible image of the fingerprint. In traditional police-related acquisition, fingerprint capture would originally have been achieved by a process of inking the fingertip and then rolling the inked finger across paper, leaving an impression of the ridge pattern behind, but in most cases nowadays, and certainly in biometrics applications, the image is acquired directly from the finger in so-called 'livescan' acquisition. Typically this requires a user to place a finger on a small flat-surface pad where the image is taken directly, but sometimes it requires sliding the finger across a sensor (this may be familiar to readers who have experienced this type of sensor built into a home computing platform, for example). Other technologies—based, for example, on capacitative sensing—are also possible.

In implementing a biometric system to analyse and compare fingerprints for identification purposes, we then need to extract information which allows us efficiently to determine a matching 'score' when compared against a pre-stored list of known individuals or when comparing two images to verify a claimed identity. The simplest way to do this is to make use of the fact that there are two very dominant aspects of ridge patterns which typically characterize fingerprints. These are illustrated in Figure 10, and are referred to as *minutiae*.

The first minutia type is a *ridge ending* which, as the name implies, is simply a point where a ridge terminates, and the second type is a *bifurcation*, which is a point where a ridge divides into two separate pathways. Each minutia can then provide several pieces of information, some or all of which can be used in determining a degree of match between two samples. Thus, each minutia can most simply be labelled with its type (ridge ending or bifurcation) and its position within the image plane (essentially recording its x–y coordinates). A *direction* (with respect to a fixed axis within the plane of the image) can also be associated with each of the minutiae.

Given the long history and ubiquity of fingerprint checking, it is not surprising that there are also other levels at which the images can be compared. For example, the ridge patterning viewed at a rather higher level generally reveals 'standard' patterns often referred to as loops, arches, whorls, and so on. While these features in themselves can be found in many individual fingerprints, they can still be useful in identification. One useful technique is first to carry out a quick examination at this high level. This might remove the need for further matching checks if, for example, one of the samples to be matched has an arch configuration while the other is predominantly characterized by a whorl. Overall, then, we can see that the features which we might use in the comparison process form a type of hierarchy, which can be exploited as we construct computer-based algorithms to compute a matching score between two fingerprints of interest. Fundamentally, however, matching is based on identifying the position and possibly the type of the set of minutiae defining the fingerprint of a particular individual. Another advantage of this approach is that it turns out we need not store very much data to represent each fingerprint—indeed, around 400 bytes seems to be adequate to encode around 100 minutiae to describe an individual fingerprint.

All this sounds extremely easy but, as we generally find in the real world, the process may not be quite as straightforward as it first seems. This is because in the real world, rather than in the realm of the theoretical, imperfections can be found and errors can occur, and Figure 11 illustrates just some of the issues which might be encountered. For example, skin is elastic and hence deforms when pressure is applied as the finger is placed on the sensor. Quite apart from the possible effect on the relative positions of the minutiae this factor, along with the accumulation of dirt, grease, sweat, and so on, or dryness of the skin, can cause some potentially serious disruption to the ridge patterning which the image captures.

The first example in Figure 11 shows an image where many ridges are obliterated, perhaps because of excessive applied pressure and/or the effects of dirt/sweat, while the next shows the effect of too little pressure being applied. The third shows what happens

11. Some captured fingerprint images. Top left: excessive pressure on the sensor, top right: too little pressure on the sensor, bottom left: incorrect placement of the finger on the sensor, bottom right: motion blur.

when the finger is incorrectly angled as it is placed on the sensor, and the fourth example shows a blurred image, most probably because of movement of the finger on the sensor. And, obviously, some individuals are likely to have damaged or worn fingerprints, perhaps because of their occupation or lifestyle (manual workers such as bricklayers, for example) and we all suffer from cuts and abrasions from time to time. For all these reasons we need both to build a degree of flexibility into the processing algorithms we use, while we can often apply helpful preprocessing to the raw image (transforming its appearance in some useful way before we attempt to analyse it) in order to rectify some of the problems encountered. A range of established *image processing* algorithms are available to help with issues such as the presence of noise in the image, improving the contrast of the image to make the ridge patterning clearer, thinning the lines corresponding to ridges in order more easily to identify the minutiae characteristics, and so on.

There are, too, some other advantages of the fingerprint as a biometric modality. The long history of the successful use of fingerprints is one obvious benefit, and experience built up through their use in forensics is extremely helpful. People generally leave their fingerprints around as they move through their environment, giving rise to *latent* fingerprints as a source of data which can be helpful in many applications. Neither should we overlook the fact that the long history of fingerprint processing means that extensive databases of fingerprints exist, and these can be extremely valuable for research and for the design of practical systems. On the more negative side, we have seen the variable quality often associated with real image capture, and the fact that physical contact is still usually required to capture fingerprint information can lead to some resistance because of hygiene concerns. This requirement for physical contact also, at the least, limits the distance over which acquisition can be achieved. Nevertheless, this is a widespread, popular, and successful modality in practice.

Biometric systems based on iris patterns

When light enters the eye, it is focused by the biological lens on an area at the back of the eye (the *retina*) which contains physiological light-sensitive sensors (*receptors*, the so-called *rods* and *cones*) which register the incidence of the illumination and accordingly generate electrical signals which are transmitted through various layers of processing cells (generally abstracting a higher level of information at each stage), until the effect of the incident light is interpreted at a high-level centre in the brain, giving us the sensation of visual perception and an understanding of the external object from which the light originated.

However, as the light enters the eye, its intensity is regulated by means of a filtering system. If the incident light is very intense the filter effectively reduces the amount of light transmitted through the lens, while under low levels of illumination, more of the incident light is allowed to penetrate to the lens. This filtering mechanism allows us to cope with a visual world in which we cannot always choose the level of illumination available. We are all familiar with the appearance of the human eye, so most people will be aware that the *pupil*, at the centre of the eye, is the aperture through which the light enters, and that the coloured disc which surrounds the pupil is called the *iris*. It is the iris which, because it can elastically expand and contract, acts as the filter which automatically adjusts (by means of the so-called *pupillary reflex*) to the ambient lighting level, opening up to allow more light through the pupil or closing down to restrict light entry depending on circumstances. These immediately externally visible parts of the eye are shown in Figure 12, where the characteristic patterning of the iris is clearly apparent.

From a biometrics perspective, however, what is of primary interest is that the iris has a specific patterning, visible both during dilation or constriction, which is to all intents and purposes unique to each individual. The patterning can be seen as a variety of striations, rings, furrows, vasculature, and so on specific to each individual eye, which give it its unique appearance (even the two irises of a single individual are different), and which makes it of great interest in the context of biometric person identification. This is readily apparent in Figure 12.

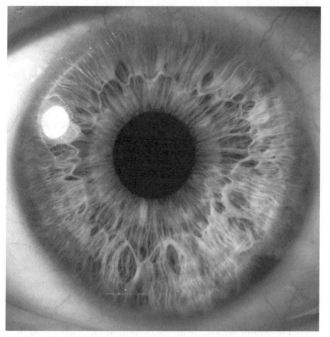

12. Image of the eye region, showing the typical patterning of the iris.

In fact, there are a number of reasons why iris patterning is very popular as a source for collecting identification information, and which provides a significant contrast to the case of the fingerprint which we previously looked at. The iris is not as immediately accessible as the fingerprint, but has an advantage that its appearance can easily be captured without direct contact between an individual and a sensor. In recent years there have also been considerable advances in systems which can capture the iris pattern at increasing distance from the subject, although most systems currently still require the user to stand at a fixed position in front of a camera, thereby providing a more standardized image and one in which the cooperation of the subject removes problems arising from different 'pose' (angle of the iris with respect to the plane of the capturing lens), obscuring of the iris, and so on. Although the iris is not immune to physical damage, because it is embedded within the eye and protected by the *cornea*, a membrane which covers the iris area, it is much less susceptible to routine or inadvertent damage which is more easily inflicted on the fingerprint, especially in some occupation-related situations. There are also other iris properties which are of particular interest in a slightly different context, to which we shall return in Chapter 4.

Because the iris image is generally also captured by a simple camera without the constraint of having direct contact with the sensor, the images collected can be subject to greater variation than we would expect with some other modalities (the fingerprint is again an obvious example here, although we saw that the fingerprint image is itself subject to degradation processes not experienced with the iris). I have already mentioned pose—the way the eye is presented to the imaging device—while positioning within the field of view of the camera is also likely to be variable. Illumination is important, since it is necessary to extract the iris part of the wider eye area from the overall captured image before we can extract the features necessary to attempt identification, and we must remember that the eye is intermittently obscured by the eyelid as we blink, occluding part or all of the iris itself. And we should not overlook the fact that the dilation or constriction of the pupil will affect the amount of iris patterning visible in the captured image (if the pupil is fully dilated, for example, this automatically reduces the amount of the iris patterning visible). The process of separating the iris from the overall image of the eye which the sensor captures is called *segmentation*, and a considerable amount of research has been invested in developing reliable segmentation algorithms, which are such a fundamental operation in the processing chain required to obtain the required biometric data.

Having identified and isolated the iris area, we then need to extract the characteristics/features of the patterning which we can use to build a representation of an individual iris, which can in turn be used to define a standard biometric model to compare either against validated examples taken from a specific individual (in a verification scenario) or against a set of labelled samples taken from a range of known individuals (recognition scenario). In the case of the iris, the process of extracting relevant features is less immediately obvious than identifying simple minutiae in the fingerprint. Although there are various ways in which this can be achieved, the most common approach has been to use a mathematical process which, expressed simply, explores the changes in the patterning across the iris using a method which measures modulations in the texture of the iris (colour is thought to be potentially less stable and is generally not considered), and can be made invariant to size fluctuations in the image. A key idea in the basic formulation of the process is then to transform the mathematical representation of this patterning to a normalized string of binary (0/1) digits which can be easily compared between irises. This standard representation is often referred to as the *IrisCode*, and consists of a 256-byte string of zeros and ones, this string being an encoded representation of the unique iris patterning of each individual eye.

We can see that if our iris model is a simple string of binary digits, comparing two irises (the important 'matching' process) becomes relatively simple, because we just need to compare the two strings of interest bit-by-bit, noting the number of times the corresponding bits in the two compared images are different. This number of differences can be seen as a simple measure of how dissimilar the two irises are. A large number of differences will suggest the two irises come from different eyes, while a small number of differences (the variability implicit in the process makes it, to all intents and purposes, impossible that we will extract two identical IrisCodes) suggests a much greater likelihood that the two images being compared come from the same iris. In the now familiar way, we can set a threshold to make a decision about whether to accept or reject the hypothesis that the two images come from the same physical iris.

Very positive claims have been made about the robustness of iris pattern comparisons as a biometric measure. Although it has been estimated that the likelihood of IrisCodes from different irises exactly matching by chance is phenomenally small (at around 1 in 10^{50}), we should nevertheless recognize that, as with all practical implementations, there are potential sources of difficulty which arise in a practical scenario, leading to some distortions in the matching process. Examples of some of the issues which can affect the captured image (and therefore either prevent a good image being obtained or, at least, meaning that the IrisCodes generated become distorted), include issues such as eyelashes projecting on to the iris, obscuring its patterning, or eyelids occluding the required image, and we have already noted the possibility of pose with respect to the camera affecting the nature of the image. The wearing of contact lenses, and indeed ordinary glasses, can cause unwanted reflections which might distort the information captured, while motion blur during capture can cause severe problems. While many of these issues can be largely overcome by retaking the image (i.e. allowing a subject more than one attempt to demonstrate an acceptable match), this needs to be controlled carefully, and it is important to be aware of the existence of potential sources of error in practical situations, as is the case for all biometric modalities.

Overall, however, iris-based biometric systems are generally regarded as highly reliable. This is a modality which clearly has a benefit of requiring no direct contact from the user, although obtaining a good image (because of difficulties in aligning the eye to the camera) has traditionally been cited as a significant challenge (an issue which has become much less problematic in recent years as techniques and technologies for image capture have improved). But a major benefit is that the capture technology is relatively simple. It also has the advantage that, with an appropriate set-up, there is a possibility of deploying iris-based identification in a covert capture scenario if required. We will return to some of the other properties of this modality in Chapter 4 when we explore the potential vulnerability to deliberate attack of a biometrics-based system.

Facial images in biometric recognition

It is not surprising that using images of the face to recognize individuals has attracted much attention over many years, since this is perhaps the principal way in which we, as human beings, generally recognize others (the sound of the human voice is comparably effective in the human repertoire of recognition feats, but even this probably cannot compete with using the visual appearance of the face). We usually give this little thought, accustomed as we are to operating within an environment where the visual sense is so important, yet if we do stop and consider this activity, we will see how remarkable a facility is our capacity to recognize, and distinguish between, different faces.

It is so remarkable largely because each human face is, at a high level at least, very similar, and what distinguishes one face from another essentially comes down to rather small variations in specific landmarks. Moreover, we seem to be pretty good at this task even when part of the face is occluded, because of a hairstyle or the wearing of a hat/head-covering, for example, or when—as happens all the time—the facial expression changes to reflect mood or current circumstances, or when we see the face from an angle which changes its detailed appearance. We can usually distinguish between family members, even when striking family likenesses have been passed along the generations. Other factors can also play a part in changing facial appearance, but maybe even more subtly. For example, some of us wear glasses (but not necessarily all the time), but maybe we change their style from time to time. Make-up is often applied to the facial area or, after a holiday, a suntan may change the colour and texture of the skin, and we can think of many other influences on the appearance of the face, yet most of us seem able to cope with such a range of natural variability, usually—forgiving the pun—without batting an eyelid. This familiarity with facial recognition, and the ease with which we are able to carry it out, has meant that even before biometrics became the pervasive technology we currently know, computer scientists, engineers, forensic scientists, and others have worked to develop powerful algorithms capable of extracting useful information from facial images.

In fact, although here we are mainly concerned with the face as a means of identifying its 'owner', the face can reveal information about a surprising range of characteristics, including for example the age of the subject; whether the subject is male or female; the subject's mood (happy, sad, etc.); the subject's current state of mind (depressed, anxious, etc.); the degree of engagement of the subject with the viewer; the level of familiarity of the viewer to the subject, and so on. And, perhaps surprisingly, all these properties can be discerned on the basis of often very small changes in appearance. Little wonder, then, that the face has a unique place in our understanding of human behaviour, human psychology, and human communication, let alone as a basis for identifying individuals.

The appearance of the face is determined by its underlying skeleton, and by the hard/soft tissue, muscles, and skin which overlay this. At the outer layer we also observe differences of skin colour and texture, and then, obviously, we find specific landmarks, such as the eyes, nose, mouth, ears, and so on. All of these contributing factors can form the basis of extracting features to describe and, eventually, recognize an individual face. And then we must be aware that (much more so than with many other sources of biometric information) the properties which are observable are likely to change as we age. Thus, when we adopt a biometric system, we need to be aware of these factors and, especially if we want a system to operate for an extended time, try to endow the system with a capacity to ignore or adapt to these changes, in just the same way as we humans seem well able to do.

If we propose to adopt the same sort of approach to face recognition that we have been using as a general principle for biometric systems, then this brief discussion should also give us some clues about what to look for. So, measurements of characteristics around the eyes, nose, mouth, eyebrows, and so on, might be a good place to start (and this means, inevitably, that our algorithms will have to embody a means of identifying these 'regions of interest' within the overall face image). Indeed, in many applications we have to begin by latching on to the face from a generalized image which may contain a complex background, or where more than one face occurs—the *face detection* problem, as it is known—although very effective algorithms for this can now be readily found.

The sorts of features which can then be extracted and used for recognition processing include, among many others, the following examples: the position of the corners of the eyes; the distance between eye centres; the position/dimensions of the tip of the chin; the position/distance between mouth extremities; information about lip edges and contours; the relative positions of some/all of these landmarks, and others. And then we can start to consider more subtle and/or more complex properties such as the following: quantitative measurements of the curvature of lips, eyebrows, etc.; measures relating to skin texture; appearance and texture of the hair; and more complex shape descriptors to describe key landmarks. If we approach face recognition in this way, it is easy to see that we will end up with a set of defined features which will allow us to apply the sort of standard techniques for achieving recognition which we have encountered previously with other modalities.

In fact, there are a number of other techniques which have also successfully been adopted for face recognition. Although the details are outside the scope of this book, one well-developed idea begins with facial images in their raw form starting at the pixel level and generating a more compact representation without having to extract individual features. This statistical process begins with a training set of representative faces and transforms this set of images into a much reduced set of characteristics which are nevertheless the main elements which have made up the faces which constitute this training set. A specific face can then be represented, in essence, by a weighted combination of these main elements—that is, any particular face can be seen as being made up of these elements in different degrees, as it were. When trying to recognize a specific face, the input face can be positioned in the same reduced dimensionality 'space' and assigned to the individual whose known face (as a designated user of the proposed system) is closest to the questioned image. This approach has been found to be fast and tolerant to natural variability in individual facial appearance.

Because face recognition is one of the most powerful recognition modalities immediately available to humans, there is a huge literature, especially within the field of experimental psychology, which reports studies of the capacity of humans to recognize faces. A familiarity with this type of work has often provided computer scientists and engineers designing automated biometric systems with a better understanding of some of the important processes which underpin facial image analysis, and has also directly influenced the development of algorithms for practical application.

Overall, then, automatic face recognition offers the major advantages of ready public acceptance while being particularly unobtrusive and entirely familiar and non-threatening to users. It provides contactless acquisition of relevant data and, as may be important in some applications, can be used covertly as well as overtly. An additional positive aspect is that there is also the possibility of using a face recognition system with pre-existing 'legacy' images (photographs, for example), which broadens the scope of application further. On the other hand, facial image processing is easily degraded because of issues with pose (the orientation of the face to the capture device) and the high potential for occlusion of important features, and is quite sensitive to environmental conditions such as illumination. Also, non-cooperative subjects, especially in non-covert scenarios, can challenge the effectiveness of the capture and processing stages, and there is also a threat to operational robustness as a result of spoofing (trying to fool the system by using artefacts), but more will be said about this in Chapter 4. All of these issues have stimulated a huge amount of excellent research, contributing both to rapid practical improvements in recent years, and an ongoing programme of development of very interesting and powerful emerging biometrics-based techniques.

The handwritten signature

Some modalities depend on human action to make biometric data available, and the handwritten signature is a good example of this. The fact that handwriting involves a complex interaction of mental processing and (extremely complex) muscular activity, requiring both coarse and fine control mechanisms, suggests intuitively that handwriting is likely to be a potentially rich source of information about an individual, and this is borne out by our experience of the considerable variability in writing styles between one person and another. Writing, and a search for the factors which define writing style, also has a very long history, and there is now a very extensive literature on this topic. Indeed, the notion of developing an objective description of writing style has played a large part in the context of forensic analysis and criminal investigation for many years, principally, it should be said, using the skill of human observers with experience and specialized analytical skills.

Of particular interest in the field of biometrics is a specific form of handwriting. The handwritten *signature*, by its very nature—the execution of a signature is generally a much practised and regularly repeated action, leading to the development of a sort of internalized 'programme' which we execute without much obvious conscious thought—is a piece of handwriting which is unique to each individual. It is also the case that the individual signature has an important function in many practical situations, especially some of those which relate to authorization processes, where our signature signifies our approval for something to happen, and is taken as confirming our identity in relation to that authorization process. We also use the signature as a means of individual identification in less critical transactions—signing a note or a letter, for example, and it is therefore not surprising that processes for objective analysis of an individual's signature and a need to establish its authenticity are still important tasks in forensic applications. Increasingly, however, with the spread of biometrics-based applications, the use of the signature for identification purposes is now moving into more routine and everyday applications. Consequently more and more of us will be aware of using the handwritten signature as a means of securing access to platforms such as laptops or mobile phones, and in many other areas.

The adventurous reader may like to undertake a simple experiment, which is to ask as many acquaintances as possible to donate a sample of their usual signature (alternatively, samples can readily be found in a simple internet search—including samples from many famous people). A study of these will quickly reveal a huge variety of signatures styles. Sometimes the signature is a simple handwritten version of the 'owner's' name. Sometimes it is a much more free-flowing, embellished, and approximated execution of the name, while in other cases the signature becomes a highly stylized and much more symbolic mark. But perhaps the most common style—and this seems often to be the style developed by well-known 'celebrities' who frequently are asked for an autograph (just watch the end of any professional tennis match!)—is more of a rapid flourish or a kind of pictogram, which often bears no easily identifiable relation to the name it signifies.

This notion of 'style' of signature perhaps explains why the handwritten signature is potentially such a useful source of biometric data—it generally takes a form which is highly individual, is easy to generate and entirely non-invasive to capture, is non-controversial as an individual identifier (since everyone already uses it), and, as a bonus, has a well-established legal status. On the downside, however, because its form results from a specific individual *action*, it only exists when this action is executed, and there is considerable scope for a degree of variability in its form, since this will be dictated by a number of unpredictable factors, such as the signing environment (think about signing for a parcel delivery on your doorstep, balancing a package in one hand and scribbling with an unfamiliar stylus on a surface held by another person!), the signer's physical and mental state, and so on. We should not forget either that we leave samples of our signature around all the time and these, unlike fingerprints or, even more so, iris patterns, are likely to be easier to reproduce by people other than the genuine signer under certain circumstances. As we noted in Chapter 1, we refer to this type of modality as a *behavioural biometric* which, for the reasons given earlier, has a number of implications for the way in which we process and use the data generated.

The signature is in some ways more complex than some other common modalities, and this is not just because of the potential for significant variability already noted, but also because it offers different strategies for data acquisition. Most obviously, the signature can be seen as a simple two-dimensional image, and we can thus capture its form using standard camera-based imaging techniques. This will allow us to extract from the captured image a wide range of defining features (dimensions, shapes, measures of curvature, information about ascending and descending segments with respect to a defined baseline, and so on). We often refer to such features as *static* features, and they are available to us at the point of signing but also, if necessary, at other times and places by extraction from a carrier such as a letter,

Making biometrics work

some formal document, or whatever substrate on which the signature is found, by a similar imaging process. A potential weakness of such flexibility, however, is that the information which we thus use to process and establish the veracity of the signature is exactly that which is also available to anyone who has previously had sight of the form of the signature.

An alternative approach is to capture not (or not only) the visual appearance of the signature itself, but the pattern of pen movements which was executed in order to write the signature. This allows us access to a much richer source of information which is not available merely by observing the final image of the completed signature. While it is possible, to some extent, to access this type of data via a simple camera, the process is not easy or particularly accurate, and thus has very considerable limitations. A better, and more usual method, is to utilize a special-purpose device (most often an electronic tablet and pen), whereby the movement of the pen across the surface of the tablet can be tracked electronically, usually by sampling the position of the pen tip at regular intervals within a two-dimensional x–y coordinate frame. This process results in the signature representation becoming a long sequence of data packets which are each defined by a time stamp, and the horizontal pen location and the vertical pen location at that time instant.

From this, the 'image' of the signature can be generated if necessary, but it is also now possible to measure many more individual characteristics of the signing process itself. For example, because we can now measure time intervals and spatial distances, we can measure pen speed, either overall or, more importantly, because we can now segment the signature into individual strokes, at the stroke level. We can also measure pen acceleration, patterns of pen stops and starts, and large volumes of data which tell us about how the signature was executed (effectively, although indirectly, about the individual muscle activity patterns underlying the signing process). Most importantly, we can determine information about the *sequencing* of actions, such as the order in which strokes are made. Furthermore, with the right acquisition hardware, we can also measure information about the pressure of the pen on the writing surface, and how this changes, we can note pen tilt and rotation about various axes, and so on. All of this can be added to each data packet and, because it is not visible to an observer only of the completed signature, is much more difficult to reproduce, thereby increasing the security of the whole biometric identification process.

Features of this type are all collectively referred to as *dynamic* features, in contrast to the static features described at the beginning of this section, since these features carry time-related information, giving much more valuable personally unique information than many of the static features. It is also important to note that static features are generally readily recoverable from dynamic capture, while the reverse is much more difficult and often impossible (we sometimes refer to 'pseudo-dynamic' features, where we have been able to infer dynamic information from static appearance, but these usually provide a very limited form of dynamic input).

It is not surprising, then, that dynamic capture is preferred whenever possible, but it is also very useful to know that valuable individual characteristics can still be obtained even where only static capture can be accomplished. Still, the literature shows that improved accuracy, reliability, and robustness can generally be achieved using a dynamic approach while, at the same time, the less accessible information afforded by this means also imposes a greater degree of difficulty for a forger to achieve successful imitation of another's signature.

Signatures are particularly sensitive to the conditions under which an 'attack' (i.e. efforts to obtain information to help with fraudulent signature generation) is attempted. Some examples include so-called *passive* (zero-effort) forgery, *shot-in-the-dark*

Making biometrics work

imitation, *knowledge-based* attack, and *shoulder-surfing* observation. Passive forgeries, such as the coincidental similarity of simple signatures of common names like 'John Smith' discussed in Chapter 2, could often be considered, to all intents and purposes, not to be attacks at all, but nevertheless can give rise to errors and operational inaccuracies in some systems. These are *passive* forgeries, since they usually are not intended to deceive.

Shot-in-the-dark attacks tend to be opportunistic, and occur when an attacker is aware of the name of the subject of the attack, and simply signs using a handwritten version of that name. While, obviously, this is likely more often than not to be unsuccessful, there is also the likelihood that on occasions this type of attack will succeed, especially if the name is a relatively simple one or a commonly occurring one, or both.

In a knowledge-based attack, the attacker has some real knowledge of the form (and maybe inferred information about the execution pattern) of the signature which is being attacked. This might have been acquired from sight of a document or other 'carrier' of the signature (most of us still carry around debit/credit cards which handily provide a visible sample of our signature!). Of course, as we have discussed, this type of knowledge may get the attacker only so far—because the sample will reveal only static information (and, possibly, a small amount of inferred or pseudo-dynamic information), and this is not likely to be very effective if the imitated signature is presented to a biometric system which uses dynamic analysis.

Finally, shoulder-surfing refers to a situation where a potential attacker manages to contrive an opportunity to observe the genuine signer in the act of signing (surreptitiously 'looking over the shoulder' of the signer, for example). It is then possible, again perhaps depending on the nature and complexity of the signature, to glean some idea both of the visual appearance of the genuine signature and also learn something about the act of signing which the genuine signer adopts. In principle at least, this might possibly lead to an attacker making a successful attempt to fool a system which utilizes some elements of dynamic representation.

In thinking about these issues it is important to remember some of the basic principles we considered in Chapter 2. A system can be set up to require much better or more accurate signature attempts (generating higher matching scores) to be provided if successful recognition/verification is to be declared, or can conversely allow for considerable variability to be accepted (lower matching scores) for an acceptable match to be declared. It will be apparent that achieving the right balance can be especially tricky when we are dealing with behavioural biometrics such as the signature, where natural variations are implicit and unavoidable in the act of signing.

In fact, it is known that some individuals have developed a signature which is simply very variable, irrespective of obvious problem areas such as the signing environment, hand/arm injuries, and so on. Some people just have a signature form which is basically unstable. As we saw in Chapter 2, we often refer to such system users as 'goats', and they do present difficulties in practice. We can either adjust the operational parameters to avoid them being routinely (falsely) rejected by a system, or we can increase overall security for the majority, but at the expense of making system interaction very inconvenient and ineffective for these individuals, who will often find themselves locked out. Sometimes we may have to exclude goats from a system altogether.

Overall, then, the handwritten signature, while offering both advantages and disadvantages, both for the user and a system operator (as is the case with all biometric modalities, in fact), has some particular characteristics which make it a very interesting modality to study, and recent years have seen a better understanding of the problems associated with this modality and substantial improvements in the capabilities of biometric systems based on signature processing.

Modality diversity and selection options

It should now be more evident how different modalities exploit fundamentally different types of human characteristic. For example, we have looked at a modality which uses a familiar and easily accessible physiological characteristic, which is readily visible externally and where the features of interest are easily seen, easy to understand, and, to an extent, a direct imitation of what human experts are capable of. This, the fingerprint modality, has been in use for many years, stretching way back before the advent of automated processing. We have also examined another physiological biometric modality, this time based on iris patterning, where the source of the relevant information is again readily accessible, but where the characteristics which are used in processing are rather less immediately obvious. Additionally, in the first case, acquisition of the data usually imposes some limits on the proximity of the user to the system (and in most current systems requires direct contact with a sensor), although these limitations are reducing as technology develops to find alternative means of acquiring the features of interest. In the second case, no physical contact is required, and current technology is allowing capture of iris data at increasing distances.

The third modality we have examined, based on capturing facial features, is perhaps the most familiar of all, and aligns directly with how humans most commonly identify each other in a remarkably efficient and flexible way. The fourth modality we have considered, the analysis of the handwritten signature, is fundamentally different insofar as the biometric data in this case do not exist unless and until a specific action is carried out by the user and, as we have seen, even then the relevant information may or may not be easily observable directly, depending on the way in which the sample for the source data is captured. We have also seen that both the variability of samples, and opportunities for malicious attack, are somewhat less predictable for this behavioural modality.

This brief modality-focused discussion (and there are many more modalities we could have explored if space permitted) illustrates quite neatly a fundamental issue relating to the adoption of biometric technologies. All modalities offer both advantages and disadvantages, and performance will depend very much on the nature of the proposed application, the nature of the population who will be the primary users of the system, the environment in which it is to operate, and many other factors. It is not possible to identify a universal candidate for the 'best' modality, since this can only be determined when all the factors are known and have been carefully considered. What we can see, however, is that by understanding that the implementation of all the available modalities, different though they may be, is based on a common set of principles, it is easy to identify the key elements which we have to consider, and we know that the underlying operational issues will remain pretty much the same.

Armed as we now are with an understanding of the basics, some specific information about individual modalities, and thus a good working knowledge of biometric systems in general, we are in a strong position to move on in Chapter 4 to consider both some ways in which we might be able to achieve improvements in performance and, on a related matter, to look more carefully at some potential areas of vulnerability for biometric systems and see how we might improve their resistance both to erroneous performance as a result of natural occurrences and also when subject to malicious or fraudulent attack.

Chapter 4
Enhancing biometric processing

Can we improve system performance?

Having grasped the basics, in this chapter we will explore how the field of biometrics is developing, and the main ideas promoting the improvement of the accuracy, reliability, and effectiveness of biometric systems. Some of the basic concepts introduced here are already playing a part in real system implementation, while others remain as options for consideration in specific applications or operating environments.

It is unlikely that any individual biometric modality operating alone will completely meet all the desirable criteria for a given task, especially when we consider the variety of issues which need to be considered in any practical situation, such as the performance and accuracy achievable, acceptability to the intended population of users, convenience in use, optimality for the envisaged application, and how well matched to operational environment is a particular chosen set of system parameters. And, conceivably, with relatively little thought, we could come up with a list of various other important factors too.

A general approach for addressing these issues would be to imagine an implementation which aims to integrate evidence about identity from more than one source. This should allow us to achieve greater *accuracy* (several pieces of independent evidence all pointing us to a robust decision), but also provide the option, in some circumstances, of offering an element of *choice* to the user.

This leads to the notion of *multibiometrics*, which can be embodied in a system in various different forms. Some possibilities include the following. We could, for example, adopt a configuration which, while using data from the same biometric source, subsequently processes this in multiple different ways, and then combines the results. This might most typically therefore involve a single modality, but a multiple processor system, most readily embodied in what is usually designated a *multiclassifier* system, where we process the same data using different classifiers, each of which manifests individual strengths and weaknesses, contributing in different ways to an overall classification decision.

Alternatively, we could look at something rather different, such as a structure involving a greater degree of what we can refer to as *integrated decision-making*, seeking independent evidence from different biometric modalities, and then combining this evidence in reaching an overall decision. A further benefit of using more than one modality is that this can also introduce an element of choice into the use of a system, offering options for an individual user to choose a preferred modality or, indeed, avoid a modality which may be difficult for a particular individual to use. Such a configuration can be considered an example of what we generally term *multimodal* systems, for obvious reasons.

Finally, we can adopt what I will call an *extended* system configuration. In this structure, we can think about adding further information, not necessarily conventional biometric information, to enhance decision-making. The most common configuration in this category is one which uses so-called *soft biometrics* as a supplementary evidential source. Let's take a brief look at each of these different options in turn.

Adding extra power using a multiclassifier configuration

We have seen how at the heart of a biometric system is the notion of extracting features from a biometric sample to characterize the unique identity of an individual, and we know that, provided these features have been well chosen, their values will vary only to a limited extent across genuine samples from a given person, while showing much greater differences across samples from different individuals. We also know that the two most fundamental operations for a biometric system can be summarized as follows: first, to construct a model of an individual based on a statistical analysis of the variability of the features derived from specific samples of known provenance (the *training* samples); and second, in operational mode, to receive a set of feature values from an individual user sample and to *classify* this set of values as matching the model of a particular individual.

A problem which obviously can arise is, not surprisingly, that the design of the classifier can draw upon a wide range of different approaches, each of which may result in a different classification performance. Each classifier is likely to give a different level of performance, depending on many factors such as, obviously, the operating principles of the classification strategy, but also the statistical make-up of the samples used, the nature of the features which have been selected, the operational parameters set into the system, and so on. So while we can try to 'optimize' the design of a particular type of classifier, we also know that a different classifier might perform better or worse if we were to make a substitution, but we cannot necessarily predict whether in the long term, given that the samples it will have to process are entirely unknown, we have made the ideal choice for all circumstances.

One way to address this potential difficulty is to try for the best of all worlds. Specifically, we can throw into the mix more than one classifier, and try to use the accumulated 'knowledge' of a *set* of classifiers working together, rather than relying on a single option, thereby—we hope—gaining the benefit of greater power overall in our decision-making, avoiding errors that might be attributable to one particular classifier, and providing more resistance against weaknesses in any specific approach to classifier design. We generally refer to this approach as using a *multiclassifier* configuration, and the general idea is to exploit the strengths of the component classifiers, while mitigating their individual weaknesses.

This raises a new issue, however, since we have to find a way to combine the evidence of identity provided by each separate classifier. We also have to remember that the classifier can either produce a 'score', signifying a degree of match with the statistical model of an individual enrolled in the system ('The score for this sample is X') which would then need to be translated into a 'hard' decision about which individual is identified on the basis of the sample presented, or the system can make for itself this specific decision based on a threshold of similarity which has been set ('This sample is from person X'). So we have options at least about whether to use scores or hard decisions at the individual classifier level prior to coming to an overall decision (and, indeed, other options are also available). This can lead to a variety of strategies for how to combine this identity evidence. Briefly, options range from the obvious and very simple approach of allowing each individual classifier to make its own identification decision, and then accept, as it were, a majority verdict, through an alternative approach which combines individual scores in some way, to a configuration where we use specially designed combination approaches which introduce a greater degree of 'intelligence' into the process, involving the ability for classifiers to 'negotiate' with each other based on their individual detailed assessments, leading to an agreed final decision.

If we choose the individual component classifiers with care (specifically, if we can choose a set of classifiers the individual components of which generate largely non-overlapping errors), then we will usually find that we can improve the recognition performance achieved compared with what is possible with any individual classifier operating alone. Thus, without in any way changing what is required of the user, but only changing the internal system-processing structure, we have an opportunity to improve the accuracy of system performance. While such a system is likely to be considerably more complex than using a single classifier, its new structure should perform better but also be transparent to the user.

Increasing flexibility using multimodal systems

In this approach we aim to adopt more than one biometric modality in order to obtain broader evidence about individual identity. For example, we could ask a system user to provide both a fingerprint sample *and* execute a sample signature before trying to establish identity or check a claimed identity. Or we could use a combination of, say, a facial image and an iris pattern, since both require the user only to look at a camera, so this might be considered more convenient, or we could use three different modalities, or whatever we choose.

We would then have information about identity based on a wider range of evidence but, as in the previous case of the multiclassifier processing structure, we would need to combine these pieces of evidence to make a decision about the identity of the information provider. Again we can consider a range of options, the most obvious of which is perhaps simply taking all the features extracted from all the modalities used, and stringing all these together as a single data source for processing (feature-level information fusion). We could instead allow each modality to reach its own decision and then combine the information at this level (decision-level information fusion). Consider a basic verification task as an example: we can ask each modality to reach an accept/reject decision on the validity of the claimed identity, which then leads to a range of other options, such as (a) a sample must be accepted in all modalities to be accepted overall, (b) a sample must be accepted in a majority of modalities to be accepted overall, or (c) a sample must pass in K out of N modalities to be accepted overall. In fact, the first two options are specific cases of the more general third option. We might also consider a further configuration, where we use the scores generated in each modality and combine these in a way which gives greater weight to some modalities than to others (score-weighted information fusion).

However we choose to combine the different sources of evidence, though, we can expect some advantages to accrue by broadening the base of the identity evidence acquired in this way. First, we are exploiting more information than would be available from a single modality, and thus it is intuitive to suppose that the accuracy with which we achieve identification should be improved. Second, we can now make the system inherently more resistant to attack, since an attacker will have to find a way of breaking more than one modality in order to fool the system into an incorrect overall decision which is to the impostor's advantage. Third, and a benefit which is often overlooked, is that by making available a range of modalities, we can allow a degree of choice to the user in relation to which modalities (s)he uses, if this is beneficial.

A user may feel uncomfortable because giving a fingerprint sample requires him to touch a surface which appears dirty. Another user knows that she has a naturally highly variable signature, which often causes irritating false rejection. Someone who has lost his voice because of a medical condition, either temporarily or permanently, may find it impossible to use a voice-based system. And we can imagine many other scenarios where using a particular modality is either impractical or undesirable from the point of view of a user. Adopting a multimodal configuration offers a viable way of handling these cases or simply just providing a better experience for a user. Indeed, we can see that a multimodal system provides a direct option for many cases of what is sometimes referred to as *exception handling*—a situation where, for one reason or another, a user would not be able to use a particular biometric system, perhaps because of some physical impairment, or simply because of personal preferences. Thus, adopting this type of configuration can in principle greatly enhance the power and applicability of the introduction of biometrics into practical operational scenarios.

Although providing many advantages, we should also be aware that adopting a multimodal configuration will also carry some disadvantages. Apart from the added complexity of the processing algorithm which will be required, a system which requires the provision of infrastructure to collect data in several modalities will also obviously add to the basic set-up costs and also incur greater operational overheads. Not only that, there will be a time penalty incurred in requiring users to provide multiple different samples, especially if the collection apparatus has to be physically dispersed. This will inevitably increase the time required for system use on the part of the user, leading, at best, to a potential perception of inconvenience or, at worst, the time required will exceed what can reasonably be allowed in practical terms (consider, for example, the queue at airport immigration control). Nevertheless, where the nature of the application permits it, and where a viable system can be engineered, this option can offer obvious advantages. This last question of *usability*, however, is a crucial one here (and is a factor which is often not taken fully into account), and a proper consideration of this aspect of system design is very important in weighing up the advantages and disadvantages in a given application scenario.

Soft biometrics as additional identity evidence

It may have occurred to some readers already that we have so far ignored a whole area of information relevant to describing the identity of individuals. When we give information about ourselves in signing up to some new service, facility, or entitlement, we are generally asked for information which quite clearly says something about our identity, but which has nothing to do with fingerprints or irises. I am referring, of course, to questions we are frequently asked about characteristics such as our age, our gender, height, or other distinguishing features we may exhibit. Characteristics such as these certainly contribute to our identity, yet are not used as the foundation of any biometric modality. This is because these characteristics are not unique, but are shared by a (usually large) number of others as well. However, if you know my age, you have immediately narrowed my identity down to a much smaller population of possible individuals than if you did not have this information. Such characteristics are often referred to as *soft biometrics*, and have the property that they provide some (maybe important) information about my identity, but are not *unique* to me (uniqueness, it will be recalled from Chapter 1, was one of the key criteria for defining biometric data).

Soft biometrics are therefore of more than passing interest to those working in the biometrics community since it is possible, and can be very advantageous, to exploit soft biometric data to enhance the usual biometric information adopted in practical biometric systems. This type of information is also of increasing interest in other ways too, as we shall see in due course.

Consider subject age as an illustrative example. This is one of the most powerful sources of this type of information. The problem when working with age is that age is not a simple 'measurement', but is naturally progressive and continuous, changing minute by minute, day by day, so we are immediately confronted with the question of the degree of 'granularity' which it is appropriate to adopt in defining the age of a particular individual.

Rather than use a highly specific age tag, such as '28' or '42 years and 4 months' or whatever, it is more customary to consider each individual as falling into an *age band*, which is much broader than this. This still allows for many options, but one useful way of defining age using this approach is to adopt bands which represent respectively 'young', 'medium/middle age', or 'older/ elderly'. While, obviously, we can further divide these very broad classes, they are at least a useful starting point, since in very general terms they define age categories between which the most marked changes can often occur. For example, handwriting style is generally still developing while we are young, settles down considerably as we get older, but then can start to exhibit changes as we become elderly and the degree of muscular control we can exercise begins to deteriorate.

We will come back to soft biometrics in Chapters 5 and 6, but here we can briefly consider how we might, at least in a simple way, exploit knowledge of subject age to enhance our attempts to identify individuals using a slightly extended biometric system. I can illustrate this with data from experiments carried out in my own research laboratory. In a small-scale experiment in person identification, based on features extracted from the handwritten signature, we have observed, even with a very simple processing configuration, improvements of between 2 per cent and 3 per cent in error rates (depending on the classifier used) when augmenting the biometric data with age information, and somewhat more than this if we use handedness (whether the signer is right- or left-handed) as the soft biometric characteristic.

So far in this chapter we have explored some of the ways in which we might be able to extend the basic structure of a biometric system in order to enhance its performance, primarily from the point of view of the accuracy (and, to an extent, operational convenience to the user) achievable in determining identity. However, the effectiveness of a system depends on more than how accurately it can perform, since a biometric system is often deployed where identity checking is of crucial importance, and sometimes where the consequences of an incorrect decision could be very serious, and this brings us to another area in which system design has been developing in recent years.

Resistance to 'spoofing' (presentation) attacks

Given that biometric systems are often deployed to identify individuals in order to protect unauthorized access to physical spaces, virtual spaces, or important data, it is hardly surprising that such systems can become the targets of activity aimed at overcoming the protection they provide, in order to access places and/or data sources for fraudulent purposes. Thus, security issues are also extremely important aspects of strategies for the safe and robust deployment of biometric techniques.

In fact, the subject of system security is such a broad and important one that it is a topic ideally requiring a whole book in its own right. For our present purposes, suffice it to say that a biometric system has a number of potential vulnerabilities, mostly well documented, and that a considerable amount of work has been carried out over the years to ensure that practical biometric systems avoid or mitigate the effects of 'attack' by a third party. Nevertheless, no system can be guaranteed to be completely free of the possibility of attack, and therefore it is useful to consider the sorts of measures which can be introduced in order to increase security and provide protection in these circumstances. We will look specifically at two sources of vulnerability.

To illustrate an obvious potential area of vulnerability, we might consider a telephone-accessed bank account, whereby I can use my mobile phone to call up, check details of my personal account, and make transactions. If my bank account uses voice-based biometric protection (in other words, if I am recognized when I try to access the account by means of algorithms which recognize my voice characteristics to ensure that it really is me calling up), then there is a considerable motivation for criminals to find a way of bypassing this form of identity checking, either by pretending to be me in some way or by using a more indirect means of fooling the system into believing that *they* are *me*.

How might this be achieved? Some ideas which spring immediately to mind are, first, for an attacker to imitate my voice. This suggests that the algorithms used to analyse my voice should exploit a deep knowledge of detailed voice characteristics. Constructing tools for the analysis of speech signals is a well-established area of research which is highly developed, but clearly offers two different approaches in this context. First, recognition of my voice is likely to be achieved more reliably if it is known in advance what words or phrases I am actually going to use (so that the system can be tuned specifically to the way my vocal characteristics are manifest in these specific utterances), while there are likely to be greater security benefits if an attacker does not know ahead of a proposed attack what the words are which I am going to be required to speak. A second attack strategy might be to attempt what we usually call a 'replay attack', using a recording of my actual voice (if this can be obtained—an increasingly easy thing to do with current technology) to ensure that the detailed characteristics of my real voice are those which are supplied to the biometric analysis system.

This brief illustration immediately suggests not only some obvious forms which a system attack can take, but perhaps also reveals ways in which we might be able to protect a system. Let's take a look at two approaches which might be considered, especially when we are faced with the possibility of such a 'presentation attack', where the attacker aims to fool the system sensor by means of a 'spoof' input. These can be summarized broadly as follows: first, adopting physiological 'liveness detection', where the input data source is checked directly as the biometric data are acquired, to make sure that the signal corresponds to a source which can be shown to be part of a living person (as opposed, for example, to a good-quality pre-obtained artefact such as a photographic image (for, say, fingerprint or iris modalities) or a recording (for, say, voice), which would be inanimate). Second, we could use what we could call a 'challenge/response' approach, where the individual user is explicitly asked to provide information which cannot easily be predicted in advance (such as speaking a word or phrase presented at random in the speech recognition case, so precluding the use of a pre-prepared input which is not genuine).

Liveness detection is, in fact, a very generally applicable approach, but can make use of a particularly interesting mechanism when we consider the iris modality as an example. It has been shown that some systems can be vulnerable to attack when a high-quality photograph is shown to the camera (perhaps held in front of the fraudster's eye) instead of presenting the real iris itself to the capture camera, or utilizing a contact lens with an engineered iris pattern embedded. In order to illustrate in more detail some options available here, we need briefly to look again at some specific aspects of the physiology of the iris.

There are two properties of the iris which are relevant here. First, and most obviously, in order to function correctly as a means of regulating the amount of light entering the eye, the iris is subject to a reflex action in response to varying degrees of incident illumination, dilating or contracting the pupil as necessary. So, if the amount of incident light increases, the iris starts to close down, to decrease the size of the pupil while, conversely, a reduction in incident light results in pupil dilation. This activity is entirely involuntary. One check to determine whether an imaged iris is 'live' is therefore to change the amount of incident light and check pupil size changes (this is easily done because it involves only detecting the boundaries of the iris, which is an operation required in detecting the iris in the first place) and measuring their relative diameters.

A second interesting and useful property of the iris is a little less obvious. This is the fact that, again in an involuntary way, the iris is in constant (low-level) motion, resulting in continuous small fluctuations at the pupil boundaries. This phenomenon is known as *hippus* and, again, this can easily be detected when the iris image is captured at the sensor. Since the detection of both the pupillary reflex and hippus are based on involuntary but constrained and predictable activity, and especially since the required processing is relatively straightforward, this provides an obvious way to detect liveness and thereby defeat attacks which rely specifically on inanimate artefacts to try to gain unauthorized access to space or data protected by a biometric system. Similar mechanisms can be found for other modalities (for example, the fingerprint is extracted from a part of the body which can easily be checked for blood flow, or the presence of a pulse, or, indeed, can be imaged at a lower level to identify tiny pores not easily visible on the surface, and so on). Challenge/response methods, on the other hand, are well suited to, say, voice-based biometric systems or other behavioural modalities, as we have already seen.

Biometric data integrity

Before we move on, it is worth briefly mentioning another source of vulnerability, and one which could have far-reaching consequences. This arises because of the uniqueness of biometric data, since if, in the process of transmitting, accessing, storing, and manipulating biometric samples from an individual, the primary data sample is intercepted or accessed in an unauthorized way, then that biometric information is forever compromised.

While in a more general (non-biometric) context, a data sample can often be cancelled (or 'revoked', to use the common expression), this is difficult with most biometric samples since, for example, our iris pattern is what it is, and we cannot choose another one if a real sample is fraudulently intercepted. A well-researched approach to dealing with this sort of data compromise is based on what has become known as *revocable biometrics* (or *cancellable biometrics*). This approach uses the power of mathematical transformations to modify the raw captured data, so that this crucial information is not generally manipulated directly but, instead, only a modified form of the data is available. Specifically, we can make use of a particular type of transformation which operates on the raw data, but for which a reverse transform cannot be found (or is so difficult to determine as to be effectively unobtainable).

So, the raw biometric data are captured but immediately transformed so as to exist only in the transformed form and, providing the adopted transform is chosen appropriately, this new representation of the biometric sample can subsequently be used for all the operations in the typical biometric processing chain with which we are already familiar. The great advantage of this approach is that, in effect, it makes the raw biometric sample replaceable, in the following sense. Since we no longer work with the critically sensitive original data, in the event of compromise of the working sample, we can withdraw (revoke/cancel) that sample and subsequently generate an entirely new reference sample by going back to the raw data and applying a different transform. This obviously avoids compromise/interception of the raw biometric information itself, and allows us to protect this by accessing it only in the event of a problem with its current transformed version.

Of course, in order to make this a viable option for use in a biometric system, certain conditions have to be met in respect of the data transformation adopted. For example, as we have seen, the transform has to be, to all intents and purposes, unidirectional, so that the raw data cannot be obtained by reversing the initial transformation. Other conditions must also be met—for example, if two raw samples generate a match score greater than a given threshold value in their original form, then a matching score greater than the threshold must also be generated when the two samples in their transformed versions are compared, and similarly for non-matches between samples. Perhaps more subtly, there must be no match (i.e. the match score must be less than the threshold of acceptability) when seeking to match the transformed sample with its original form, and so on. However, the research literature shows that this type of approach can in principle be effective and, indeed, a number of variations on this approach have also emerged, offering another weapon in the fight to maximize the security and reliability of biometric systems.

As a footnote to this section, it is worth noting the special case of certain behavioural biometric modalities, where the nature of the sample is actually under the control of the individual owner of the biometric data in question. The obvious example is to consider the handwritten signature again. Although the original form of this biometric identifier is generally the result of development of a repeatable pattern of execution over time, in the event of compromise, it is perfectly possible, in principle, for the individual to 'revoke' this biometric data by the simple expedient of forming an entirely new signature for future use—since this is an identifier which is not irrevocably embedded in the physiology of the writer. This is an example of what we have called *natural revocability*, since it is entirely within the control of the owner of the biometric identifier in question, and requires no intervention from externally imposed techniques, but only the cooperation of the user.

Of course, doing this immediately raises some interesting questions such as: 'Is it possible to adopt a new signature which is suitable as a unique identifier subsequently?', 'Will a newly adopted signature become "stable" in the sense that it eventually becomes as readily reproducible as the signer's original (and now discarded) version?', 'How long will it take to achieve an appropriate degree of stability?', 'Are intrinsic characteristics of the original signature inevitably carried over into the new version, immediately raising its level of vulnerability?', and so on.

These are intriguing questions, but some work in my own laboratory has shown that this sort of approach appears to be a viable option, and thus 'natural revocability' can be considered as a further option in our box of tools to resist biometric system attack. It is also worth noting that there is a strong link here

between the biometrics field and the science of forensics, where
analysis of handwriting has always been a major area of interest.

Extending the application domains for biometrics-based processing

By now, it will be apparent that we have many tools and
techniques already available to help us address some of the issues
which might concern us as we see biometric systems increasingly
being introduced into everyday applications which we all have to
use. The reach of biometrics is increasing, and is likely to continue
to do so in the future. So let's now take a look at how ideas of
identity and its determination can be extended into related, but
rather different, applications.

We start from the observation that the fundamental goal of
biometrics is to make a *prediction* about an individual. So far,
this has been entirely concerned with predicting the identity of
an individual or predicting how likely it is that an individual is,
in fact, the person he or she claims to be. But is there any reason
why the data captured in a biometric system could not be used for
other sorts of prediction too? In seeking some insights into this
question in Chapter 5, we will see not only how biometrics can
offer broader options than we have so far considered, but also that
biometrics is part of a continuous spectrum of disciplines which
are interconnected, and which can be mutually supportive.

Chapter 5
Predictive biometrics

We saw in Chapter 4 that the concept of soft biometrics, which includes characteristics such as the age or gender of a subject, is a straightforward variation on the principal theme of the study of biometrics, which is to use measurable characteristics of individuals to determine or confirm their identity. Such characteristics are not unique but can place an individual within a reduced subset of a target population. This in itself could, in many applications (searching a watch list, perhaps), be a useful thing to do.

The way we have considered using soft biometric information so far is to use this as additional evidence about the identity of a person, and we have seen how exploiting this type of data can improve our ability to make an accurate identification. It is also worth asking, however, whether we could turn the issue around and, instead of supplementing conventional data with soft biometric data, try to work from conventional biometric data to predict some soft biometric characteristics of an individual. In other words, why not explore the possibility of predicting a person's age, for example, by using conventional biometric measures? There are many applications where this type of exercise could be very useful—various entitlements which concern individuals depend on their age, the obvious examples being access to particular websites, to entertainment activities, and so on.

More broadly, the availability of many state-administered social security benefits may be age-related, as is access to some physical spaces (bars and clubs, for example). The same arguments could be easily applied also to situations in which gender or other human characteristics are important in access-limiting. Indeed, there are many situations in which, even when not talking about restricting access to place or data, knowledge of such properties could be useful or important. Investigating crime is an obvious example.

Work on predicting such individual properties, the field of what we might call *predictive biometrics*, has expanded considerably in recent years, and the two characteristics most frequently studied are probably those which have already been mentioned, namely age and gender. And such predictive work is of interest to forensic scientists too. If, for example, we can accurately predict the age of a suspect from the way he walks, then perhaps even a brief or low-quality image captured on a surveillance camera might provide valuable information in the investigation of a crime. Similarly, analysis of a signature on a will or other document might be instrumental in revealing an attempt to defraud or, conversely, to support a claim of innocence. This highlights an increasing commonality of purpose between the biometrics community and those working in forensics in the past few years, which has stimulated a range of new and exciting research activities.

Let's begin by considering how well such prediction might work in practice, and we will consider just a couple of examples from the recent research literature. Perhaps not surprisingly, the largest volume of work reported on the predictive properties of biometric data (in the sense of predicting soft biometric traits) has been based on the face modality, but interest in this area has been rapidly spreading to other modalities in the more recent past, and we will take a lesser-investigated modality to use for our example here.

Predicting age from biometric iris data

Predicting age is a challenging problem, not least because, as we noted in Chapter 4, age is a continuous variable, so we have to have some notion of what degree of granularity is required in a given application. In other words, the requirement is typically to predict an age band or age range for an individual, rather than strict absolute age. However, the greater the number of allowed age bands which are defined, the less accurate is the prediction likely to become (since, in the limit, we would of course end up trying to predict absolute age). On the other hand, using too few allowable age bands will naturally reduce the value of the prediction, even though accuracy is likely to increase. At a minimum, defining three age bands, corresponding broadly to what we might call 'younger', 'middle-age', and 'older', is perhaps the most common categorization found in the literature currently. In the work referred to here, these bands are defined as *less than 25 years, between 25 and 60 years*, and *greater than 60 years* respectively.

Not surprisingly, achievable predictive accuracy, even having fixed these age bands, will vary depending on a number of other factors, such as the type of classifier used, and so on. There are also issues around the actual data used for experimentation (ensuring that enough subjects are available to make the experiments statistically meaningful, that a representative cross-section of typical users is adopted, and so on). Experimentally, using a database collected under controlled conditions, and including iris data from 210 subjects with ages from 18 to 73 years, it has been shown that an accuracy of a little over 70 per cent can be achieved using a modest number of relatively standard features extracted from the iris images. The features used can be divided into two different types, the first describing principally the geometric features of the iris, while the second corresponds to features which largely provide descriptions of the textural properties of the iris. If we test the predictive accuracy of a system using the two different feature types separately, we find that the textural features generally provide better predictive accuracy than the geometric features (although a small number of exceptions can be found, depending on other factors), while utilizing all the available features always produces the highest accuracy of prediction, whatever classifier is used. It is also possible to show that using a multiclassifier configuration, especially where the combination technique is chosen carefully, can improve performance further, generating accuracy levels of well over 80 per cent using the data available.

This could suggest that, at least in the longer term, adopting age prediction as a first step before implementing a recognition phase could potentially result in a more efficient overall process, since the age prediction can then be used as a sort of filter, reducing significantly the number of matching operations required to be carried out (probably by a factor of around 3 in this example) in a subsequent attempt to identify the queried individual.

Predicting gender from iris images

Again using the iris modality, it is possible to find reported work which aims to predict gender from the biometric data. If we base our study on assuming that only two possible genders are to be found (the lack of available data precludes any other more subtle option at present), then this should be a less complex procedure than dealing with age, where a wide range of different age bands could reasonably be considered, as we have discussed.

Two important studies have been reported at the time of writing. As we have seen, the precise performance figures are found to vary considerably depending on the exact processing configuration adopted, but the first of these studies shows that choosing the best classifier generates performance figures approaching 80 per cent accuracy when an appropriate selection of features is carried out. The second study compares a range of classifiers but also explores the comparative performance achieved when using (a) features relating to iris texture and (b) features relating to the geometric properties of the iris, showing a considerable improvement in predictive performance when textural features are adopted (typically showing around a 10 per cent improvement when using textural features, for almost all classifiers) and a further significant improvement if all the available features are used, achieving marginally over 80 per cent accuracy in the best case. This study also investigated the use of powerful, more 'intelligent' classifiers in a multiclassifier configuration when, by deploying an optimal combination algorithm performance is shown to improve prediction accuracy to a level approaching 90 per cent. However, it is somewhat dangerous, as always, to make direct comparisons, since the data used to establish these respective performance figures differs in the two studies, and thus like-for-like comparisons are not possible.

Nevertheless, it is evident that diversifying the use of biometric data to achieve the prediction of a range of characteristics of an individual appears to be possible, with a degree of accuracy that suggests exploitation of such techniques can be worthwhile in practical applications. These studies also neatly illustrate how different configurations, with differing conditions and operating parameters, can deliver different performance levels, showing how important a careful design process can be in matching configuration to application.

Extending predictive capabilities

It is now clear that, while remaining in the realms of using conventional biometric data, it is not too difficult to turn around our earlier objective and predict soft biometric data rather than simply using predetermined knowledge of this information to enhance identification performance. But the most recent work in this area has taken the idea of predictive biometrics further still, with studies which aim to predict not just the common soft biometric characteristics such as age and gender (and, of course, identity itself) from conventional biometric measurements, but which have extended predictive capabilities to so-called 'higher-level' individual characteristics, such as those which reflect an individual's mental or emotional state.

Here, for example, are some performance figures which predict whether a person is broadly 'happy' or 'sad', purely on the basis of capturing simple biometric data. In this case, we have extracted features from handwriting samples of individuals (exactly the same features as those we have previously used to *identify* individuals from their handwriting activity) who, in the experiment referred to here, had also revealed to researchers information about how they were feeling. In order to extract biometric features on the basis of which to predict emotional state, subjects were asked to carry out various writing tasks, with their writing captured using a digitizing tablet, as already described. Some of these tasks were simple copying tasks based on words/word groupings (such as copying a long sentence which was constructed so as to ensure the inclusion of all the most common letter groupings in the English language), where all subjects were required to write the same, predetermined, target text. Other tasks were 'free-form', with subjects asked to write, in their own words, a short description of a visual scene presented to them (although this is obviously not the only task which could be used and, indeed, in a practical situation, there may be no choice about how the sample is obtained).

It has been found possible to predict the subjects who regarded themselves as 'happy' with an accuracy approaching 80 per cent although, not surprisingly, the predictive capability again varies depending on a number of factors, including which task was being undertaken, the type of classifier being used, and so on. There are also notable differences, not unexpectedly, in the power of different features to contribute to the prediction process. As we might expect, the use of dynamic features of handwriting (see Chapter 3) generally produces better performance than the static features. This is likely to be a particular issue in some forensic applications, where we might usually expect only static features to be available (because forensic investigations are often likely to use, of necessity, legacy documents rather than capturing data online). However, it is also possible to show how increasing the number and diversity of available features improves performance, and with an appropriate feature set it is possible to predict emotional state with an accuracy of around 70 per cent even where only static features are available. This work, a research study carried out with colleagues in my own Research Group, represents only a preliminary study, yet offers considerable scope for further development in the future.

As we have noted, this is still a relatively new area into which the established techniques of biometrics are beginning to take us. For this reason, while there are obviously a number of important questions still to be answered, and aspects of this work still to be investigated (some of these pretty fundamental, about how we assess such predictions and how we measure degree of happiness in the first place, for instance) there are some encouraging signs here that there is more useful and accessible information embedded in biometric data than we might at first think.

The question of the prediction of the characteristics of individuals from physiological or behavioural measurements is at the heart of what biometric systems aim to achieve. While the most fundamental characteristic to be predicted is still generally the identity of an individual—a characteristic by definition unique to every person—it is evident that predicting other characteristics is also possible, offering information which can be extremely valuable in a variety of applications, even where the characteristic predicted is not unique, but shared among a group of individuals.

These recent studies provide a glimpse of some of the possible ways in which the field might develop in the future. In Chapter 6 we will conclude by looking briefly at some other ways in which the field of biometrics is developing, and indulge in a little crystal ball gazing.

Chapter 6
Where are we going?

In Chapters 1–5, we have moved from the basic principles of biometrics, and how biometric systems have developed as powerful tools for establishing or confirming individual identity, through a discussion of the principal sources of useful biometric information, to a consideration of some of the ways in which biometric systems are increasingly being made more accurate, robust, and reliable in practical applications. We have also glimpsed some new directions in which the field is moving, broadening the reach and scope of biometrics as well as diversifying the important application domains where biometric techniques can make a valuable contribution. This final chapter picks up these latter themes.

To give a flavour of how the field might develop in the future, we will focus on a selection of topics of particular current and future interest. These will include a brief look at some biometric modalities which we have not so far considered, including some perhaps surprising sources of biometric data which might achieve a higher profile in the future, some further ways in which systems can become more easily embedded in regularly used applications and a more natural part of everyday life, and some of the issues which might affect the long-term reliability of practical systems.

Expanding the range of modalities

I wrote at the start of the book of the range of possible modalities which can be used as the basis of identifying individuals, but we looked in detail at just four of these, to give a sense of the sort of information required in order to capture the individuality of each human subject. These particular modalities were chosen because they illustrate the broad spectrum of possibilities. Facial features are extremely familiar, easily accessible, and are used in everyday life by other humans to recognize individuals; the fingerprint is less easy for humans to recognize without special training, but has been used in identification tasks for well over 100 years, and has been automated for a considerable time; iris patterning is easy to understand, but is less easily accessible technically while having acquired a reputation for high reliability when suitable capture conditions can be arranged; and, finally, the handwritten signature is an obvious example of a behavioural biometric source, where an individual needs to carry out some specific action in order for the required information to emerge. We have also looked at the use of voice characteristics. Intuitively, we all know how relatively easy it can be to recognize a person from vocal information, even on a telephone line, where the frequency range of what we are receiving is severely reduced. At the time of writing, voice-based biometric recognition is gaining a high profile through its introduction on a large scale in personal online banking applications, to give just one example.

However, a brief tour of a few other modalities will fill out the picture of the variety and diversity of what biometric information has been used, what can be used, and what might be used in the future, though this list will still not cover all the many possibilities.

Hand-based biometrics

Hand shape and hand geometry can be used as a biometric identifier. This data source has a long history, and is notable, if not for the highest accuracy, at least for the ease of use it offers, its general robustness, and the fact that it can be installed in operating environments where some other systems might be unsuitable.

It turns out that the hand shape of any particular person is very individual. Typically, the user of a system adopting this modality is asked to place her hand on a flat surface, above which a simple camera system is able to capture an image of the hand. The platen on which the hand is placed is often fitted with physical positioning pegs, which allows the user to slide the hand into a standard position with the pegs anchoring the fingers, helping to obtain consistent positioning and thereby reducing the amount of image normalization which is subsequently required. The finger joints, for example, and the points marking the extremities of each finger form a set of natural markers, from which can be derived a set of specific measurements to characterize each individual. Thus, a set of measurements which describe finger thickness, width, and length, distances between joints, and other relative dimensions form the basis of a useful characterization of individuals by capturing a representation which satisfies the uniqueness criterion sufficiently to function as a viable biometric modality. Indeed, the basic technique is relatively simple in terms of its required technology, very easy and natural to use, and offers a degree of accuracy acceptable for a number of applications, especially where the number of enrolees can be constrained. On the other hand (forgive the pun!), hands change size over time, most rapidly during childhood and in the transition from childhood to adulthood, while with age hand shape can change significantly as a result of accidents or normal physiological changes associated with conditions such as arthritis and similar conditions. The need for regular re-enrolment could therefore be an important consideration.

Fingerprints are also hand-based biometrics, and so indeed are palmprints, the markings formed on the palm of the hand, similar to the patterning at the fingertips which define conventional fingerprints. Palmprints also therefore provide a source of biometric data not dissimilar in processing terms to fingerprints, offering a further biometric modality for consideration.

One further small observation: the fact that there is a group of biometric modalities all based on measurements of hand properties of one sort or another suggests that in principle it should be possible, given an appropriate acquisition set-up, to collect several different biometrics at roughly the same time, and with minimal inconvenience to the user. This could be especially useful in a multibiometrics-based configuration.

Keystroke dynamics

One consequence of the social changes brought about by technological development is that for a number of years now, many people have spent increasing amounts of time typing information into computing devices via a keyboard. Although even here technology continues to develop, resulting in touch-sensitive interfaces, either structured as a familiar keypad or even through capture and interpretation of freehand writing, the standard and very familiar computer keyboard is still a predominant means by which an individual communicates with computing platforms. As with other frequently adopted and repetitive actions, the typing patterns we develop, especially in relation to tasks such as typing our computer username or password, become highly predictable and repeatable.

For this reason, it is not surprising that these typing patterns have long been known to offer another specification for a biometric modality, often referred to as *keystroke dynamics*. In this case, a major advantage is that the identification process itself can be closely bound to the usual and expected activity of a user since if, for example, the typing of the password is taken as the source of biometric data, the usual log-on procedure can be integrated with a biometric check of the identity of the system user. During the execution of the tracked input typing sequence it is not difficult to collect, for example, timing information which, together with the actual alphanumeric sequence typed, robustly characterizes the user. The basic measurements often used for this purpose are, first, *inter-key timings* (the time elapsed between key presses for successive typed characters) and, second, *key hold timings* (the elapsed time between each key press and subsequent key release for each typed character).

This sort of information has been found effective even for short target strings of characters, maybe a password of just seven or eight characters, although the technique can be extended beyond passwords to other target character strings, and indeed, to strings of any length. With small user populations this modality has been found useful, and new developments in this area are still emerging, despite the long history of this modality, although this is currently not a widely adopted source of biometric data in practice. We can see that this is another example of a behavioural modality, since it requires a specific action/behavioural pattern to be generated by the user in order to obtain relevant biometric data.

Ear shape

Ear shape has emerged relatively recently as another physiological biometric (even though the identifying properties of the ear have been known for more than a century), since the ear has a unique appearance for every individual. There are various ways in which the shape and appearance of the human ear can be characterized for identification purposes. The most obvious way is to identify the major landmarks on the ear (defining the shape and size of the various cavities, ridges, and so on) and then extracting measurements relating to these landmarks and their relative positions. Another interesting approach which has been suggested is to consider the image of the ear at the most basic digitized level, where each point (pixel) in the digitized image is considered individually and a model evolved in which it is considered that each point attracts each other point with a strength which is related to their relative intensity and (inversely) to the distance between them, with the overall image then modelled as an 'energy field', with peaks, troughs, and ridges which can provide the basis for identification measurements. A disadvantage of using this modality is that, as with some others, the ear is subject to problematic imaging because of issues of pose and especially the high risk of occlusion because of hairstyle, let alone head coverings and other sources of distortion.

There is nevertheless some evidence that this is a promising direction for the future, and that using the physical properties of the ear is a potentially useful modality to add to the toolbox available to the designer of biometric systems, but it is a relatively new modality and one not yet fully tested in practical applications.

ECG and EEG measurements for biometrics

It is perhaps difficult to imagine anything more personal in relation to the human body, the source of any biometric measurement, than the human heart and the brain. Both organs generate continuous electrical activity, the heart because of its muscular movement and electrical control system, and the brain because it constantly generates electrical impulses in many millions of neurons. In each case, it is possible to detect this electrical activity at the body surface using electrodes attached to the body at appropriate points. And because we are all physiologically different, these signals generate different activity patterns in each individual, thus making them potential candidates for biometric identification tasks.

If we look first at the heart, it is customary to determine its activity by means of an electrocardiogram (ECG), in which the electrical activity at each heartbeat is detected by electrodes placed on the body surface at standardized locations. Specifically, the ECG can be seen as a repeating signal with the characteristic overall shape familiar to most readers, but the precise form of this repeating pattern is determined by the physical features of each individual heart, different in each of us. Moreover, this type of signal is easily analysed using well-known mathematical techniques to extract features which characterize its appearance with a high degree of accuracy.

It is hardly surprising then that the ECG should be investigated as a possible source of identification data at the individual level, and there is an increasing body of work which has studied the ECG as a biometric data source. However, there are some obvious questions to be asked in particular about the capture of the required data for use in a practical scenario. To use this modality requires that subjects are willing to have skin electrodes attached, although this is not an uncomfortable or invasive procedure. Because of the obvious medical applications of ECG capture, there is also much commercial interest in developing improved schemes for data acquisition—using wireless transmission of acquired data, for example. The rate at which the standard cyclic signal pattern repeats is variable (the heart beats at a different rate in different people), but while this is sometimes the result of individual differences, and therefore a useful biometric characteristic, it is sometimes a result of changes brought about by stress, physical exertion, and so on, not to mention changes indicative of disease.

So while clearly a good candidate for providing the basis of a new and potentially powerful biometric modality, utilization of ECG measurements is not necessarily as straightforward as we might like. A particularly useful feature, however, is that because the body generates ECG signals regularly and continuously (typically around between 60 and 80 times each minute), and because it is an involuntary activity, this type of modality can easily provide continuous monitoring of identity, unlike many other modalities where a specific act of data collection is required each time an identity check is undertaken. There is now a solid body of work which has been reported on adopting the ECG in biometrics, and work continues to bring this closer to viability as a suitable modality for appropriate applications.

Further complexities arise when the electroencephalogram or EEG—the pattern arising from electrical signals in the brain—is considered. Again, the EEG has been shown to embody very individual characteristics, but the capture of EEG signals in a convenient and consistent way, and the variability of the signals in relation to the activity of the subject, introduce significant potential difficulties, particularly in environments where the conditions cannot easily be controlled.

The capture of EEG traces requires that electrodes are positioned on the head. In medical applications, the number of electrodes used can be significant and, although when used for biometric identification tasks, fewer electrodes can be deployed, there are still questions about how to optimize the number and positioning of these, while potentially the additional burden on the subject and the fact that attaching electrodes is not a natural process for most people, remains a barrier to routine use, although it appears that a capture infrastructure can be embedded into familiar headgear in some circumstances, reducing this problem.

The other principal issue is that brain activity—and hence the acquired signals—changes substantially depending on what the subject is doing. For example, the differences in EEG patterns between a resting state (where no particular activity is being undertaken) and when the subject is receiving some stimulus are considerable, yet contriving a situation in practice where the capture conditions can be precisely controlled and consistently reproduced is very challenging, making comparisons between different samples potentially problematic in many practical applications.

However, reported results suggest that where conditions can be carefully regulated, impressive performance can be achieved, and an advantage is that EEG signals are very person-specific and hence should provide good quality biometric data. There is an increasing interest in this type of biometric modality and, as with the ECG, this is an area which is generating some very interesting new work. It is also true that commercial development of appropriate acquisition tools for both ECG and EEG is rapidly moving on, not just because of the potential for use in biometrics, but also because of the range of opportunities in healthcare scenarios. Moreover, a great value of these emerging modalities is that they provide inherent liveness detection and are therefore more resistant to attack than some other more traditional modalities.

The problem of ageing in the design and use of biometric systems

It is impossible to escape the fact that all human beings age, and we are all aware that the natural ageing process brings about changes in a number of aspects of both the physiology and overt physical appearance of individuals, and their capacity to carry out normal day-to-day activities. We all know, for example, that our facial appearance can change significantly as we get older, and various studies have identified the sort of things which tend to happen—the appearance of furrows and creases, a thinning of the lips, loss of muscular tone and/or a greater accumulation of fat, a tendency for skin to sag, and so on—in what can be a rather depressing way! Unfortunately, the changes induced entirely naturally through the ageing process are also those on which we often base the operation of biometric systems, and so our biometric profile can change continually throughout our lifetime. While over a short period of time this is unlikely to present a problem (since, as we have seen, no two biometric samples from the same person are ever likely to be identical anyway, and various environmental or behavioural effects can cause differences in acquired biometric patterns), there is nevertheless the potential for a problem to develop over longer time periods.

It is unfortunate that the effects of ageing occur in relation to pretty well all biometric modalities. From a physiological point of view, changes in skin properties occur as we age, resulting in poorer elasticity, and changes in moisture content, giving rise, for example, to potentially greater difficulty in making good contact between finger and sensor in fingerprint capture. Not only that, but increasing wear across the ridge patterning is likely to occur with age (especially so for individuals who have perhaps been involving in long-term rough manual activity), while age also brings an increasing likelihood that scarring or accidents affecting the fingers will have introduced spurious changes in patterning, or will have destroyed information previously present in the patterning.

As a further example, the iris modality is an interesting case, where ageing tends to have an effect via a rather different mechanism. Since the iris regulates the amount of light passing through the pupil, dilating it as incident light levels decrease, and constricting it as light levels increase, the actual amount of the iris patterning visible at the image acquisition stage is related to the incident lighting level. One of the effects of ageing is that the muscles which contract the pupil function less efficiently. This results in a tendency for less of the iris patterning to become visible as we age, and this in turn makes it more difficult reliably to extract a full range of features on which to base the identification process.

All this is particularly unfortunate because, as we get older, our memory functions also begin to deteriorate, and experimental evidence shows that our ability to remember passwords (especially where we have multiple codes to deal with) diminishes, and thus this section of the population is precisely that for which the benefits of biometrics could offer significant advantages. In other words, the fact that a major feature of biometrics-based identification is that we carry our identifiers with us, generally without any effort to remember anything, suggests that this would overcome a substantial area of difficulty which develops as we age. On the other hand, we must also remember that our cognitive skills (and perhaps our inclination to learn new activities) also diminish with ageing, and this suggests that we need also to be aware of the potentially increased cognitive overheads involved in interacting with a, probably unfamiliar, biometric system.

Early work in biometrics tended largely to ignore this issue, but now that the field has reached a level of maturity and reliability which has generated a range of practical applications, some of them with long-term implications, it is increasingly important both to understand and appropriately to manage the effects of ageing when a biometric system is deployed.

In fact, we need to be aware of two (interrelated, but not identical) issues about ageing, since in the context of biometrics, the term 'age' can refer to two different effects. Most obviously, we can consider the notion of the *chronological age* of a subject, the length of time for which a person has been alive. However, from the perspective of biometrics, we can think of age in another way. When we enrol on a biometric system, our biometric characteristics are used to construct a model, taking account of the natural variability likely to be found in different acquired samples, which form a template—this is then used as the basis for defining our identity when we present future samples to confirm who we are. This template embodies our biometric profile as it appears at the time of enrolment, which can occur at any chronological age.

However, with the natural ageing process, as time passes, our actual biometric profile will gradually change, which means that the similarity between a presented sample and the model held on the template is likely to decrease at a rate determined not just by environmental changes, but by the inherent changes which chronological ageing has brought about. We generally refer to this process as *template ageing*, since the model held on the reference template will age (become less representative of the owner) with the passage of time. In fact, from the point of view of biometrics, it is obviously template ageing which is the more critical factor but, as is intuitively apparent, how the template ages will be significantly affected by our chronological ageing as well.

This raises some obvious practical difficulties for biometric systems, since enrolling a group of users, and then freezing the template for all time, will almost certainly start to make the system less reliable and accurate as time passes. For this reason, we need to find ways to avoid the problems which therefore naturally arise over longer time periods. An obvious strategy comes to mind immediately, which would be periodically to update the template by asking users to re-enrol, thereby ensuring that we always work with a template for which the effects of physiological ageing have thus been taken into account. While clearly an effective approach in principle, in practice there are some obvious and rather severe disadvantages to this approach. First, we need to know how frequently we should do this, yet there is no established body of knowledge which can tell us what the optimal time between updates should be and, in any case, the available evidence suggests that it might be different for different modalities (and maybe for different users). So, six-month intervals might sound better than a five-year interval, but this then brings other problems, some of them immediately obvious. Re-enrolment introduces both significant inconvenience for the user (the process will almost certainly require intervention in order to validate the enrolment process, and will probably involve presenting other reliable corroborating forms of re-identification (think, for a start, of the intricacies of renewing a passport), but is also a costly option for the system operator. There are thus a number of difficulties in routinely adopting this option.

Another way of approaching the ageing issue would be to try to identify features to be extracted from biometric measurements which are not affected by ageing, or at least which are minimally changed by the ageing process. This has proved to be a rather difficult problem to crack and in practice generally, at best, buys us a little more time, without eliminating a trade-off between reliability and re-enrolment costs.

A more effective option might be to investigate the effects of ageing in different modalities and develop a more rigorous and quantitative understanding of the effects which ageing has on the measurements used for maximum identification accuracy. We can then try to use this understanding to model the way in which ageing is likely to change the individual representation held on the biometric template, and thereby predict the ways in which we should modify this as time passes. And there is also the option of making more use of multimodal systems, where we might modulate the effects of ageing in one modality by spreading the risk of significant changes across more than one modality, or where we can make selections based on our knowledge of changes occurring over time. This also links neatly with the approaches we considered in Chapter 4, where we added subject age as a factor in the identification process, though this focuses more on chronological ageing than template ageing.

This is a fascinating area, where much progress has been made in recent years, and where the importance of taking the ageing process into account when implementing practical systems is increasingly recognized.

Human interaction with biometric systems: usability

We turn now to another aspect of the implementation of biometric systems, and one which many would argue has not always been considered as carefully as it should. This centres around the question of what is often referred to as system 'usability', which is concerned with the relationship between system characteristics and what is required of the user, particularly in relation to the ease with which user and system can interact. Poor usability will inevitably lead to, at best, an increased probability of poor performance and, at worst, a resistance among the target user base to participate.

At a high level, the search for a high degree of usability perhaps intuitively begins with providing a clear understanding of the benefits of the proposed system to the user, and thus persuading her that using the system is a worthy and valuable thing to do. Beyond that, we can see that we have to achieve further targets. For example, we need to find a way to ensure that the user is provided with appropriate knowledge of how to interact with the system—what is the intended means of using it. At a lower level still, we need to consider carefully exactly what using the system entails for the user, and making sure that we design and implement the system so that interaction can easily be managed by users.

Let's consider some examples. The first issue will require, in essence, a distillation of the arguments scattered throughout this book about the nature of biometrics, assembled and tailored to the specifics of a particular application. A good example to think about might be the plan some years ago to introduce a national biometrics-based Identity Card into the UK for all its citizens. The plan was ultimately abandoned because, at a political level, although many points in favour of and against such a scheme had been put forward, the argument had been ultimately won by those who were not convinced by the arguments made about the potential benefits compared to those which raised concerns, especially in the context of civil liberties and matters of privacy. This may seem even less logical to the proponents of such a scheme now than it did at the time, given how similar national ID schemes have emerged in many other countries. But here was an obvious example of the arguments failing to persuade a majority of those concerned with decision-making of the benefits which could accrue.

At the second level mentioned above, we can perhaps think about the use of automated passport checks at airports. An early scheme involved the use of iris scan technology, but more recently a digitized facial image embedded in the passport is adopted. Anyone who has used such a system will be aware of the elaborate nature of the instructions provided, and the care taken, to ensure that each user understands what needs to be done in allowing the system to read the passport, and to collect the current image from the client. Despite this, it is far from uncommon to see users finding great difficulty in operating the system correctly.

At the third level, we may think back to the original iris-based airport passport control. I myself did not always find this easy to use—the main difficulty being one of managing to align the eye correctly at the image acquisition point in order to obtain a good iris image. This was not always easy and, for me at least, but others too, I believe, not infrequently ended in failure, and the need to reroute through the normal passport control desks.

These examples, and particularly the last one, point to a number of issues of relevance here. One issue concerns the way in which technology has been increasingly influenced by good practice in design for usability, based on a long-standing history of work in human–computer interaction (HCI) and ergonomics. In the early days of computer-based systems aimed at everyday and widespread use, there seemed generally to be an implicit acceptance that potential system users would have to adapt their behaviour to suit the way the system had been designed. However, as time has passed, and especially as users are being seen more as 'customers' rather than just users, HCI has turned the old principle on its head, to the point where there is now much more a recognition that systems must be structured and implemented so that they do the hard work of adapting to the needs of the user, rather than the other way round. In a biometrics context, we see how this has led to important work over the past few years, for example in iris recognition, where systems have been developed to make iris image capture much more flexible. So, instead of asking a user to stand at a specific point, and move around until a good camera alignment is achieved, there is an increasing move towards being able to capture an iris image at a distance, and while the user is on the move.

But there are other important issues to consider too, some of which link us back to earlier points made. For example, the physical characteristics of individuals vary considerably, partly through ageing and partly just because everyone is different. If, for example, we are using a hand-based biometric we need to be aware, first, that hand size will change as we move from childhood to adulthood, but also that even across a population entirely composed of adults, a significant variation in size and shape of individual hands will be found, and this is an issue even before we consider changes which occur because of physical illness, or the effects of accidents, and so on. Likewise, our ability to see clearly, hear well, and remember things, is a variable which depends on individual differences as well as being affected by age. Designing for usability has to deal with these factors unless we are content for some people to be excluded, and in biometrics we also have to note that it is precisely the differences between individuals which makes biometrics work in the first place. However, there is considerable evidence that in relation to security as in most aspects of life, individuals—more often than not subconsciously—carry out some rough sort of cost–benefit analysis in deciding whether to bother with a new idea or not, so designers of biometric systems will, if take-up in an increasingly security-aware society is to be encouraged, need to be more and more sensitive to good usability practice.

Past, present, and future: some reflections

The topic of biometrics has had a very long history. There has always been an interest in measuring properties of human beings (indeed, the term 'biometrics' originally referred to just such a wide remit, and there is still a degree of confusion of the use of the term in this wider context rather than its association with the more specific aim of person identification, which is the meaning we adopt here), but the biometrics field defined in the latter sense, and as used in this book, has also developed over a long period. Recognizing faces has been an interest of computer scientists and engineers for decades, while adopting fingerprints as a means of identifying people can be traced back according to some sources to the ancient civilizations of around 6000 BC, where potters would embed a fingerprint into the clay to signify their identity as the creator of an object. The use of fingerprints in criminal investigations itself began more than 100 years ago, and it is not surprising, therefore, that with the development of computer technology, automated fingerprint processing should have become an early application for biometrics, with the specific development of *automated fingerprint identification systems* from the 1960s to the present day, with widespread current deployment.

In the present, the use of biometric protection of personal devices is now also becoming more and more widespread. Smartphones and tablets are increasingly protected using fingerprint information or facial images as a means of 'locking' them. Voice recognition, signature, and even generalized gesture recognition, together with a revitalization of the possibilities of keystroke dynamics (or a popular present-day variant, monitoring touch-based patterns to take account of changing interaction modes) are all increasingly to be found in everyday situations.

It is interesting to look at various surveys of how people react to this increase in the availability and uptake of security procedures based on biometric technologies. There is ample evidence that attitudes to the use of biometric identification have changed in recent years, and that most users are relatively positive about adopting biometric technologies, but there is also evidence that many people have a poor understanding of how these technologies work (it is hoped that a book such as this one may help in this respect!). It has been found that people have an apparent preference for some modalities over others (though this may reflect availability and therefore exposure rather than inherent preferences) and that they can quite easily make an analysis which relates the potential complexity and understanding of interaction procedures to the context of application. There is, it seems, a significant concern about issues of security and privacy of data, and the development of practices which can encourage older people to become confident technology users is only just becoming a matter of significant practical concern. But there are signs that biometrics-based systems are reaching a level of maturity which is bringing with it a sense of such systems being normal and acceptable in many everyday situations.

What I hope will have become evident during our journey through this book is how the fundamental principles of operation are common across all modalities, how systems can optimally be configured, and how systems can be improved and vulnerabilities addressed. While biometric systems offer only one part of a solution to the problem of confidently identifying individuals, or ensuring that people really are who they say they are (topics such as encryption of data, for example, could fill a whole book on their own), what we can say is that the techniques of biometrics offer the opportunity to bind information to individuals in a way which more traditional approaches to authentication do not. And we should also bear the following in mind: our society has now become almost obsessed with automated interactions (how many human tellers does an average bank now employ, and how easy do you find it to pin down a human agent with whom to interact across the wide range of transactions you complete every day?). As a consequence, we seem to have to juggle with remembering passwords, either using the same password for all transactions simply for convenience (which is, of course, not best practice), or accepting that we have to remember a whole set of different passwords, maybe even a different one for each application we wish to use. This can be a challenge to everyone, but can be especially difficult for the elderly.

Not only do we therefore tend to write down our passwords, immediately potentially compromising security, but there is clear evidence that people too often opt for easy-to-remember passwords ('password' is a common password, and '1234' a common PIN, not exactly supporting a high degree of security!). Various surveys have investigated how many different passwords have to be remembered and, although different surveys come up with different answers, some have suggested that an average number is around twenty. This perhaps gives a perspective on why biometrics offers such promise, and why we find biometrics-based technologies increasingly attractive—we carry our biometrics with us, and effectively have to remember nothing.

It is striking how the deployment of biometrics in government-based applications has developed. Opportunities for beneficial adoption range from the use of biometrics in travel documents (passports, visas, and so on), through specific targeted deployment (for example, preventing 'double-dipping'—the creation of multiple identities—in relation to the uptake of entitlements such as social benefits), or the effective and secure management of inmates in prisons, through to the major national ID card programmes now in operation. We can easily think of many other potentially important operational environments—in the military sphere, for example, or in voter registration, and so on. At a more individual level, we have discussed recent developments in personal banking, and more generally in managing the security of mobile communication platforms.

Similarly, we must acknowledge that, as with any technology, biometric systems carry some small but real risks, some of which we have discussed, and we should not underestimate the importance of human and social perceptions, such as a disinclination to touch surfaces used by a multitude of different users, or a worry (however unfounded) that shining light into the eye to capture an image might be harmful, and a host of other issues which might, to a greater or lesser degree, put barriers in the way of general acceptance. And, although we have only touched on this, questions about privacy of data are a genuine concern to many people. Only further research, and making a better job of explaining the benefits of biometrics, while addressing the legitimate concerns of potential users, will promote the transition to the routine and universally accepted use of these remarkable technologies in the future.

What I hope has emerged from this book is that there is every reason to make an effort to get on board the biometrics train, and that there is nothing to fear by doing so. Knowing more about how biometric systems work can help enormously to persuade us of the benefits of this approach, and can likewise make us aware of some of the issues which will still benefit from further research. Knowledge is perhaps the greatest protection we have when we embark on the adoption of any technology. We are in a time of exciting change and development in biometrics, and improvements are rapidly working their way through from the research laboratory to the marketplace. Moreover, the framework of biometrics-based human identification is beginning to stimulate new and exciting possibilities of further useful applications in the longer term.

Biometrics is increasingly and manifestly having an impact on everyday life. Properly understood and sensitively applied, it can improve our security, increase our confidence in processes which are designed to dovetail into the way modern society works, and enhance the convenience and power of technology, for the benefit of everyone.

References

Chapter 4: Enhancing biometric processing

Abreu, M., Fairhurst, M. C., Analysing the benefits of a novel multiagent approach in a multimodal biometrics identification task, *IEEE Systems Journal*, 3, 410–17, 2009.

Erbilek, M., Fairhurst, M. C., A framework for managing ageing effects in signature biometrics, *IET Biometrics*, 1, 136–47, 2012.

Fairhurst, M. C., Erbilek, M., Analysis of physiological ageing effects in iris biometrics, *IET Computer Vision*, 5, 358–66, 2011.

Galbally, J., Gomez-Barrero, M., A review of iris anti-spoofing, *Proc. International Workshop on Biometrics and Forensics* (IWBF), 2016.

Islam, T. and Fairhurst, M. C., Natural revocability in handwritten signatures to enhance biometric security, *Proc. International Workshop on Frontiers in Handwriting Recognition* (IWFHR), 791–6, 2012.

Matsumoto, T., Gummy and conductive silicone rubber fingers, *Proc. ASIACRYPT* 02, 574–6, London, 2002.

Chapter 5: Predictive biometrics

Abreu, M., Fairhurst, M. C., Erbilek, M., Exploring gender prediction from iris biometrics, *Proc. Int. Conf. of Biometrics Special Interest Group* (BIOSIG), 2015.

Erbilek, M., Fairhurst, M. C., Abreu, M., Age prediction from iris biometrics, *Proc. IET Conf. on Imaging for Crime Detection and Prevention* (ICDP), 2013.

Fairhurst, M. C., Erbilek, M., Li, C., Study of automatic prediction of emotion from handwriting samples, *IET Biometrics*, 4, 90–7, 2015.

Fairhurst, M. C., Li, C., Da Costa-Abreu, M., Predictive biometrics: a review and analysis of predicting personal characteristics from biometric data, *IET Biometrics*, 6, 369–78, 2017.

Sgroi, A., Bowyer, K. W., Flynn, P. J., The prediction of old and young subjects from iris texture, *Proc. Int. Conf. on Biometrics* (ICB), 1–5, 2013.

Tapia, J., Perez, C., Bowyer, K., Gender classification from the same iris code used for recognition, *IEEE Trans. Information Forensics and Security*, 11, 1760–70, 2016.

Further reading

Burge, M. J., Bowyer, K. (Eds), *Handbook of iris recognition*, Springer, 2016.

Fairhurst, M. C. (Ed.), *Age factors in biometric processing*, Institution of Engineering and Technology (IET), 2013.

Guo, G., Wechsler, H. (Eds), *Mobile biometrics*, Institution of Engineering and Technology (IET), 2017.

Jain, A. K., Bolle, R., Pankanti, S. (Eds), *Biometrics: personal identification in networked society*, Kluwer Academic Publishers, 1999.

Li, S. Z, Jain, A. (Eds), *Handbook of face recognition*, Springer, 2011.

Maltoni, D., Maio, D., Jain, A., Prabhakar, S., *Handbook of fingerprint recognition*, Springer, 2009.

Marcel, S., Nixon, M., Li, S. Z. (Eds), *Handbook of biometric anti-spoofing*, Springer, 2014.

Pirlo, G., Impedovo, D., Fairhurst, M. C. (Eds), *Advances in digital handwritten signature processing*, World Scientific, 2014.

Rathgeb, C., Busch, C. (Eds), *Iris and periocular biometric recognition*, The Institution of Engineering and Technology (IET), 2017.

Ross, A. A., Nadakumar, K., Jain, A. K., *Handbook of multibiometrics*, Springer, 2006.

Vielhauer, C. (Ed.), *User-centric privacy and security in biometrics*, The Institution of Engineering and Technology (IET), 2017.

Wayman, J., Jain, A., Maltoni, D., Maio, D. (Eds), *Biometric systems: technology, design and performance evaluation*, Springer, 2005.

Index

Biometrics

SOCIAL MEDIA
Very Short Introduction

Join our community
www.oup.com/vsi

- Join us online at the official Very Short Introductions **Facebook** page.
- Access the thoughts and musings of our authors with our online **blog**.
- Sign up for our monthly **e-newsletter** to receive information on all new titles publishing that month.
- Browse the full range of Very Short Introductions online.
- Read **extracts** from the Introductions for free.
- If you are a teacher or lecturer you can order inspection copies quickly and simply via our website.

FORENSIC SCIENCE
A Very Short Introduction
Jim Fraser

In this Very Short Introduction, Jim Fraser introduces the
concept of forensic science and explains how it is used in the
investigation of crime. He begins at the crime scene itself,
explaining the principles and processes of crime scene
management. He explores how forensic scientists work; from
the reconstruction of events to laboratory examinations. He
considers the techniques they use, such as fingerprinting,
and goes on to highlight the immense impact DNA profiling
has had. Providing examples from forensic science cases in the
UK, US, and other countries, he considers the techniques and
challenges faced around the world.

> An admirable alternative to the 'CSI' science fiction
> juggernaut...Fascinating.

William Darragh, Fortean Times